Understanding Aquaculture

Understanding Aquaculture

Erin Butcher

Callisto Reference

www.callistoreference.com

Callisto Reference,
118-35 Queens Blvd., Suite 400,
Forest Hills, NY 11375, USA

Visit us on the World Wide Web at:
www.callistoreference.com

ISBN: 978-1-64116-529-7 (Hardback)

Cataloging-in-Publication Data

Understanding aquaculture / Erin Butcher.
p. cm.
Includes bibliographical references and index.
ISBN 978-1-64116-529-7
1. Aquaculture. 2. Agriculture. 3. Mariculture. I. Butcher, Erin.
SH135 .U53 2022
639.8--dc23

Table of Contents

Preface

The farming of fish, molluscs, crustaceans, aquatic plants and algae is referred to as aquaculture. It involves the cultivation of both freshwater and saltwater populations under controlled conditions. Some of the various branches of aquaculture are mariculture, fish farming, shrimp farming, algaculture and oyster farming. Mariculture refers to the aquaculture which takes place in underwater habitats and marine environments. There are diverse methods within aquaculture where farming of aquatic plants is integrated with fish farming. These methods include integrated multi-trophic aquaculture and aquaponics. The practice of integrated multi-trophic aquaculture involves the recycling of wastes from one species as the inputs for another species. This book attempts to understand the multiple branches that fall under the discipline of aquaculture and how such concepts have practical applications. Some of the diverse topics covered herein address the varied branches that fall under this category. This book will serve as a valuable source of reference for those interested in this field.

To facilitate a deeper understanding of the contents of this book a short introduction of every chapter is written below:

Chapter 1- The farming of fish, molluscs, algae, crustaceans, aquatic plants and various other organisms is known as aquaculture. It is concerned with cultivating both salt water and freshwater populations. This is an introductory chapter which will introduce briefly all the significant aspects of aquaculture.

Chapter 2- The branch of aquaculture which deals with the cultivation of marine organisms in seawater is known as mariculture. Based on the organism cultivated, there are several types of aquaculture such as fish farming and shellfish aquaculture. The topics elaborated in this chapter will help in gaining a better perspective about these types of aquaculture.

Chapter 3- The welfare of organisms in aquaculture can be affected by a range of reasons such as disease, stocking densities, parasitism and behavioural interactions. The topics elaborated in this chapter will help in gaining a better perspective about the different species which are considered important in aquaculture and the diseases which afflict them.

Chapter 4- Algaculture is a type of aquaculture that deals with the farming of species of algae. Based on their size, algae are classified as microalgae and macroalgae. Algae cultivation is used to produce various products such as algae biodiesel, bioplastics, omega-3 fatty acids etc. This chapter has been carefully written to provide an easy understanding of these varied facets of harvesting algae and seaweed.

Chapter 5- Integrated multi-trophic aquaculture is a practice in which the inputs for one species are derived from the recycled by-products from another species. It promotes economic and environmental sustainability. The diverse applications of integrated multi-trophic aquaculture in the current scenario have been thoroughly discussed in this chapter.

I would like to share the credit of this book with my editorial team who worked tirelessly on this book. I owe the completion of this book to the never-ending support of my family, who supported me throughout the project.

Erin Butcher

Chapter 1

Introduction to Aquaculture

The farming of fish, molluscs, algae, crustaceans, aquatic plants and various other organisms is known as aquaculture. It is concerned with cultivating both salt water and freshwater populations. This is an introductory chapter which will introduce briefly all the significant aspects of aquaculture.

Aquaculture or farming in water is the aquatic equivalent of agriculture or farming on land. Defined broadly, agriculture includes farming both animals (animal husbandry) and plants (agronomy, horticulture and forestry in part). Similarly, aquaculture covers the farming of both animals (including crustaceans, finfish and molluscs) and plants (including seaweeds and freshwater macrophytes). While agriculture is predominantly based on use of freshwater, aquaculture occurs in both inland (freshwater) and coastal (brackishwater, seawater) areas.

Aquaculture is the farming of aquatic organisms, including fish, molluscs, crustaceans and aquatic plants. Farming implies some form of intervention in the rearing process to enhance production, such as regular stocking, feeding, protection from predators, etc. Farming also implies individual or corporate ownership of the stock being cultivated. For statistical purposes, aquatic organisms which are harvested by an individual or corporate body which has owned them throughout their rearing period contribute to aquaculture, while aquatic organisms which are exploitable by the public as a common property resources, with or without appropriate licences, are the harvest of fisheries.

With the growth of agri-tech, modern aquaculture is able to monitor water quality, document fish behavior and manage facility logistics. With the growing popularity of Internet of Things (IoT) technology, aquaculture facilities use smart sensors and other IoT devices to examine water quality and make real-time adjustments to maintain optimal conditions for their stock. These devices can also maintain feeding schedules, improve oxygen levels and send early warnings and diagnoses to solve disease control and prevention.

Aquaculture technology has allowed the collection of large amounts of data that helps facilities reduce costs and maintain reliable environments. Fish farming facilities use mobile devices and cloud computing to monitor their stock and share information and data in real-time with company management and other farm workers. As the industry continues to grow, software companies are developing data management tools to help facilities process incoming data and assist in making business and environmental decisions.

The methods of aquaculture's farm-to-table process can differ from species to species. Generally, there are four stages of the production chain, starting in hatcheries and ending at the seafood counter in your grocery store.

Each of these stages can vary with respect to its effect on the environment and the quality and safety of the seafood they produce, which is why the Global Aquaculture Alliance administers the Best Aquaculture Practices (BAP) third-party certification program. In the past, fish farms have had issues with respect to these four aspects of aquaculture, and BAP seeks to improve the fish farming industry across the globe. This is the only certification program that covers every step of the supply chain. You can be sure your seafood was farmed responsibly if it has the BAP logo on it.

The first stage in the aquaculture production chain is the hatchery. This is where the breeding of fish, hatching of eggs and rearing of fish through the early life stages happens. Once the animals are mature enough, they are transferred to the farm, where they are grown to harvest size, using feed produced at feed mills (another stage of aquaculture). The fish are then transported to a processing facility, where they are packaged and sent to food retailers and grocery stores.

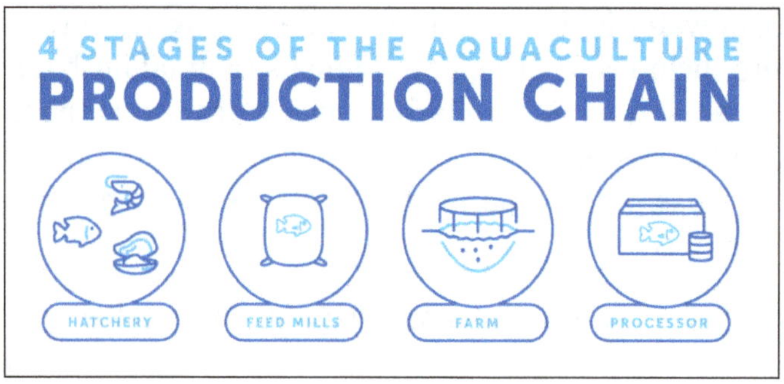

Imporatance of Agriculture

Given that overfishing of our oceans and other natural resources is continuously increasing year over year, humans need alternate sources for seafood to feed the planet's ever-growing population. "Unfortunately, the days of the ocean's natural productivity providing for the planet is over. Wild fish have been exploited for generations. Some estimate that the annual catch of edible marine protein has already passed its peak. The oceans cannot naturally provide the demand for seafood" (Positive Aquaculture Awareness). Aquaculture is the tool to fill in the gap of seafood supply. Farming fish responsibly and sustainably is the solution to providing future generations with access to healthy and environmentally friendly protein options.

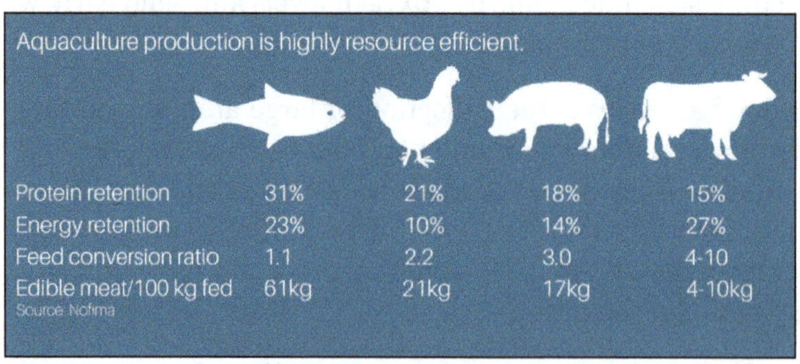

Not only is aquaculture necessary, it is also a sustainable option for consumers, especially in comparison to other farmed proteins. Seafood is highly resource efficient – it has the highest protein retention compared to chicken, pork and beef. It also has the lowest feed conversion ratio among the same forms of protein. Aquaculture has lower greenhouse gas emissions than other types of farming.

With an anticipated 10 billion people expected to inhabit the planet by 2050, the demand for animal protein will increase by 52 percent. Sustainable and healthy approaches to feeding the world are more critical than ever before. In order to sustainably feed the world's growing population with a healthy, lean protein, aquaculture's role is of the utmost importance. The primary responsibility of aquaculture is to efficiently complement wild-caught fish options to increase the amount of seafood available worldwide.

Aquaculture has the potential to improve the health of our planet and the health of our population, as long as it is done in a manner that is environmentally friendly, socially responsible, and considers food safety and animal welfare.

In some countries, aquaculture depends on capture fisheries for its supply of wild seed and feed. The production of seed from a hatchery is defined as aquaculture but the subsequent production of adult fish falls within aquaculture only if the stock is owned by an individual or corporate body during grow-out. If hatchery-produced seed are stocked in a large water body, and the production harvested by the public, it is categorised as culture-based or enhanced fisheries.

Ownership of fish-aggregating devices such as brush parks in large water bodies does not confer ownership of the capture fisheries and is not aquaculture.

Farmers have caught wild fish in ricefields since time immemorial, often building trap ponds to harvest them when water levels fall at the end of the rainy season. The fish benefit from the modified rice field which, besides providing them with a habitat, also facilitates harvest. Farmers may also make the rice field dikes and screen outlets higher to facilitate growth and harvest of wild fish without resorting to either feeding or intentionally stocking wild or hatchery-raised seed.

Stocking of water bodies with either wild or hatchery-raised seed is aquaculture if the stock is owned individually or by a corporate body until harvest but it becomes capture fisheries if there is open access for the general public. Complex social issues relating to use of traditionally communal water bodies for aquaculture lead to difficulties, often at the expense of the rural poor.

Systems and Scale

The type and scale of intensity of aquaculture systems are best considered in relation to the evolution of agriculture since agricultural and aquaculture systems follow similar paths and are often integrated. Indeed, most small-scale aquaculture may need to be integrated with agriculture, as future fish farmers already farm crops and livestock and nutrient inputs for aquaculture are most likely to originate on-farm in inland areas.

Settled agriculture was classified into three phases to illustrate the way aquaculture may evolve on small-scale farms (Ibid.), recognising that there are examples of Phase 2 which are skipped in the process of modernisation:

- Settled Agriculture Phase 1 (crop-dominated) is characteristic of pre-industrial societies with most land under food crops. Livestock are kept mainly for draught; the system includes scavenging poultry and pigs. There is limited integration between crops and animals. Large ruminants that depend on rough grazing stubble in the fields after crop harvest and are fed straw. Rice bran is fed to pigs and poultry. Livestock manure fertilises the field but the farming system is mainly crop-based due to the limited number of livestock. This system was characteristic of much of Western Europe until about 1850 and applies to many small-scale farms in developing countries today, particularly those in marginal areas. In areas of lighter population density where fish is a traditional part of the diet, farmers rely mostly on capture fisheries. As population densities increase and capture fisheries decline, there may be recourse to aquaculture. Most small-scale farmers in Asian developing countries as well as in Africa and Latin America fall within this category as they have little or no involvement in aquaculture.

- Settled Agriculture Phase 2 (integrated crop/livestock) was characteristic of much of Western Europe and the eastern USA from 1850 to 1945. These were called mixed farms because livestock production was based on arable crops and improved pasture. Livestock were closely integrated with crops because the former fed on the latter, and manure helped maintain soil fertility together with nitrogen fixing legumes. They were almost a closed self-sustaining system relying on local or farm-based nutrient cycling. Their aquaculture equivalent is the traditional Chinese integrated agriculture-aquaculture system.

- Settled Agriculture Phase 3 (industrial monoculture) is based on agro-industrial inputs. Initially developed in 1850, it only began to replace the traditional mixed farming system in Western Europe in 1950. Its major features are improved genetic varieties, chemical fertilisers, herbicides, pesticides, pharmaceutical chemicals, feed concentrates, pelleted feed and mechanisation. Most phase 3 farms raise only a single species because of increasing technical complexities and economies of scale. Much western and Japanese aquaculture belong to phase 3, as is recently developed intensive shrimp culture.

Global agriculture is divided into three classes: resource-poor agriculture; Green Revolution; and industrial agriculture. The third type of agriculture is associated with rainfed dryland, upland and swampy lowland, often peripheral areas where farming systems are much more fragile and as a consequence, complex and diverse. Abbreviated as "Third CDR" (complex, diverse, risk-prone), it broadly corresponds with Settled Agriculture Phase.

Green Revolution agriculture is usually found in national "agricultural heartlands" in fertile areas, either irrigated or rainfed lowlands, close to major population centres. Although it includes both large and small farms it essentially corresponds to Settled Agriculture Phase 3 as does the WCED industrial agriculture; both are monoculture which rely heavily on agro-industry. The latter comprises large farms and is mainly found in the developed world although there are enclaves in developing countries.

General correlation between various schema for the classification of scale of farming systems.

System			
Settled Agriculture Phase 1 (crop dominated)	Settled Agriculture Phase 2 (integrated crop/livestock)	Settled Agriculture Phase 3 (industrial monoculture)	
Resource-poor agriculture		Green revolution agriculture	Industrial agriculture
Subsistence		Artisanal	
		Specialised	Industrial
Type 1 aquaculture			
		Type 2 aquaculture	
	Balanced model		

Include sand/or refers specifically to farming systems which include aquaculture, in part or entirely.

An important difference between the types of agriculture defined by Edwards et al. is the inclusion of an integrated crop/livestock category by the former in Settled Agriculture Phase 2. This has been proposed as an important pathway by which resource-poor small-scale farming households may become aquaculturists. However, a balanced model that uses both on-farm and off-farm resources may be required for aquaculture to contribute significantly to increased welfare. Other schemes for defining scale or intensity of aquaculture systems may be related to the classification.

In a study of Francophone Sub-Saharan Africa, Lazard et al. characterised four types of aquaculture by degree of commercialisation:

- Aquaculture for subsistence (family-level);
- Artisanal aquaculture, producing for the market on a small-scale;
- Specialised aquaculture in which various stages of the production cycle are carried out by different farmers; and
- Industrial-scale aquaculture.

More recently, Martinez-Espinosa proposed two types for the development of rural aquaculture:

- Type 1 aquaculture for the "poorest of the poor" characterised by being very low cost or very low output, basically subsistence in nature but possibly bartering or selling a small part of the produce to neighbours and in local markets.
- Type 2 aquaculture which is "less poor" and comprises the operations of well-off farmers who sell most of their produce for economic benefit. This is essentially the same as the Lazard et al., artisanal aquaculture.

These categories may be compared with the more common classification of aquaculture based on productive technology, particularly feed, dividing culture systems into extensive, semi-intensive and intensive. These terms are widely use but defined in various ways, often imprecisely, or not at all. The usage outlined below is in common use:

- Extensive culture systems receive no intentional nutritional inputs but depend on natural food in the culture facility, including that brought in by water flow e.g., currents and tidal exchange.
- Semi-intensive culture systems depend largely on natural food which is increased over baseline levels by fertilisation and/or use of supplementary feed to complement natural food.
- Intensive culture systems depend on nutritionally complete diets added to the system, either fresh, wild, marine or freshwater fish, or on formulated diets, usually in dry pelleted form.

Although the classification is based on feed, increasing intensification is correlated with higher levels of other inputs such as seed, labour, capital and management. The system is less relevant for molluscs than for crustaceans and finfish because molluscs are always cultured on natural food irrespective of the level of other inputs. It has limited relevance for cultivation of aquatic plants.

Rural Aquaculture

The term rural aquaculture derives from the traditional dichotomy of development: rural or agricultural and urban or industrial. The overriding priority in rural development is to address widespread poverty and inequity in developing countries. Implicit in the term is the promotion of aquaculture systems appropriate to the resource base of small-scale farming households either through extensive or semi-intensive aquaculture. Given that the term implies an orientation towards the needs of the lower socio-economic groups, rural aquaculture may cover both inland and coastal aquaculture systems.

There is also a need to consider intensified production but in an environmentally sustainable way. Once farmers gain confidence and experience at lower levels of production, they may wish to change their level of production from subsistence or artisanal to entrepreneurial or commercial and become increasingly specialist aquaculture farmers. This requires increasing use of off-farm, agro-industrial inputs and greater involvement with urban markets. But it needs to be done in a balanced way, combining Settled Agriculture Phases 2 (integration) and 3 (agro-industrial inputs) to avoid environmental degradation.

Rural aquaculture is broadly defined as the farming of aquatic organisms by small-scale farming households or communities, usually by extensive or semi-intensive low-cost production technology appropriate to their resource base. The resource-poor base of most farms requires off-farm agro-industrial inputs to intensify production. This implies use of mainly inorganic fertilisers rather than formulated feed to provide low market value produce affordable to poor consumers.

Characteristics of Aquaculture

1. High Productivity: In comparison to agriculture or veterinary practices, aquaculture has been found to be more productive.
2. Integrated Farming: Aquaculture with agriculture, horticulture or animal husbandary is found to be more profitable,
3. Recycling: Aquaculture gives efficient means of recycling domestic and/or agriculture wastes and thus, helps in protecting our environment.
4. Rural Development: Aquaculture helps to integrate rural development by generating employment opportunities and would thus help to arrest the exodus of population from rural to urban areas.
5. Intensive Fish Culture: In aquaculture practice, owing to the fishes three-dimensional utilization of the water body can be crowded more closely (200/ m^3) and grown through water recirculation system. This gives high yield of about 25 t/ha/year.
6. Earning Foreign Exchange: Commercially important items, such as ornamental fishes, Artemia cysts, prawns, lobsters, crabs, frog legs, etc., produced through aquaculture are highly valued and can earn good foreign exchange.
7. Ranching: Fish seeds and larvae of economically important fishes of capture fisheries are artificially recruited in fish hatcheries through aquaculture, called ranching or aquarange farming. It is commonly known as "culture based capture fishery" and involves restocking of the wild stock for harvesting. This process has replenished the dwindling stock of rivers and seas.
8. Replenishment of Coral Reefs: Coral reefs that have been destroyed naturally or anthropogenically are being replenished through construction of artificial coral reef.
9. Creating Leasure-Time Activities: Leasure- time activities can be maintained through sport fishing and creating home and public aquaria.
10. Mariculture and Fisheries Enhancement: In many reef areas the use of mariculture is fast growing. It provides an alternative source of income and employment in coastal areas.

Replenishment schemes of species that has been greatly diminished has been undertaken in certain areas. This has been useful for returning species such as giant clams and trochus to reefs where they have been largely or completely exterminated by overfishing.

11. Promoting Agro-Industrial Development: Aquaculture can promote agro-industrial development through:
    - Processing and marketing of fishery products, feeds and equipments for aquaculture,
    - Seaweed culture for the production of marine celluloids, and
    - Pearl oyster culture, etc.

Types of Aquaculture

Based on Type of Water

Based on the type of water, aquaculture in three type - fresh water aquaculture where the culture take place in fresh water bodies and brackish water aquaculture, where the culture takes place in brackish water and mariculture where culture takes place in sea water. Aquaculture includes the culture of aquatic organism of food, culture to improve the natural stocks, culture of ornamental fish and sport fish, integrated farming etc.

Based on Management Intensity

Aquaculture can be categorized into five types mainly based on level of management intensity of cultural system.

1. Traditional culture: It is the simplest and an age old practice with minimum inputs managements. No selection of species, fertilization and supplementary feeding. In traditional culture system the yield is very less.
2. Extensive culture: It is an improved culture system over the cultivable species are selected and stocked species mostly depend on natural food organism. Fertilizers and supplementary feeds are used to limited extent in the properly prepared fields.
3. Semi intensive culture: It is an improved culture system over the extensive system, where the cultivable species are selected and stocked in the form of fingerling or post, larvae. Formulated compound feed as per nutritional requirements is provided. Natural food is negligible in the culture system.
4. Intensive culture: It is highly evolved culture system with more stocking density and provide more formulated feed and more aeration. Stoking is done more formulated feed and more aeration. Stocking is done with hatchery reared juveniles, water quality is maintained by frequent changing or by providing water circulation together with constant aeration.
5. Super or hyper intensive culture: The stocking density is higher than of intensive culture. Continuous water exchange is made through biological filter system. Constant aeration is

provided to meet the higher stocking densities. Encapsulated pellated feeds are provided for better growth of cultivable animals.

Based on Input Intensity

Based on level of input intensity, the aquaculture is categorized into four levels.

1. O- Level: In this level absolutely no management is found in culture system no stocking manures and supplementary feeds applied in this system.
2. I - Level: Only one management of stocking seed is prevalent. Seed is stocked in culture system. No manuring and supplementary feeding in the culture system. Only natural food organism useful as food or stocked animal is used.
3. II - Level: Two types of management of stocking seed and manure application are prevalent. Seed is stocked. Supplementary feed is not applied. Culture animals still depend on natural food organism, which can be raised by application or organic and inorganic manures in the culture system.
4. III - Level: Three types of management techniques are applied in this type of culture system. These are seed stocking, application of manured and supplementary feeding. Semi sensitive intensive and hyper intensive culture system are of this type only and the production is very high in contrast to the above levels.

Based on Water Media

Depending on the physico chemical characteristics of the water media the water are divided as fresh water, Brackish water and Marine water.

Fresh Water Aquaculture

The fresh water aquaculture deals with due the culture of the organism in fresh water resources namely rivers, streams, canals, reservoirs, anicuts tanks and ponds etc. The aspects of breeding of the parent stock growing the seed, preparing the water source for culture, stocking, water management, post stocking management methods and harvesting are included. The type of organism cultured include fishes, prawns, mussels, frogs, aquatic plants etc.

Aquaculture in Flowing Fresh Water

The flowing water are also called as lotic water. The rivers, streams, canal and distributors are the flowing water resources. The development of river fishery by auto stocking will always be taking place in rivers, streams, pools in the rivers and tanks connected to rivers by streams. The aquaculture is not possible in flowing water as the there are no control over them. The stocking of fish seed or fertilizing the water go a waste. Moreover the river will have already some fish (both predatory and non predatory). The stocking of seed in water source having fishes will not be profitable. In addition the fish seed stocked will not stay at the place of stocking. The running water are rich is dissolved oxygen. However there is limited scope for taking up fish culture in canals with less flow of water by taking up culture of fish in cages.

Aquaculture in Stagnant Water

The stagnant water are also called as lentic water. This include Lakes, reservoir, anicuts tank and ponds constructed exclusively for culture of the organism:

- The swamps are also stagnant water with less dissolved oxygen content.
- The lakes are naturally formed water bodies due to the low lying areas.
- The reservoirs are man made water bodies constructed across a river of stream.
- They are mostly perineal. The excess water is discharged through the spillways.

The tanks are perineal (water retained throughout the year) long seasonal (water retained for 8 to 11 months) and seasonal (water retained for less than 8 months). In the short seasonal tanks the fish growing is not profitable. The ponds are water bodies constructed exclusively with definite shape and size to take up culture of the organism.

In the earlier period, the man had constructed water impound in streams and rives are called them as tanks and distributaries to the paddy fields. Thus the water are stored for certain period. Aquaculture is a secondary objective in irrigation resources.

Brackish Water Aquaculture

The brackish water are areas of confluences of fresh water and sea water and the pH ranges from 5 to 27 ppt. The brackish water aquaculture is also known as Coastal aquaculture. The brackish water fish such as mullet and other fishes were cultured off the Italian coast by Romans long ago.

Estuaries back water creeks and lagoons are the main non stagnant brackish water. In these water fishes and the seed of milk fish (Chanos) mullet (Mugli Cepahlus), Elops, Megalops polynemus, Lates, Etroplus, tilapia shrimp are available. The word shrimp is used for the prawns available in the marine and brackish water sources while the term prawn is used for them in fresh water sources.

These following water are useful for collection of fish seed or prawns larvae for growing in Brackish water fish ponds. Directly the fish culture in these water are not be possible, but under favourable conditions the culture in cages or pens maybe tried. Now the collection of prawn larvae and selling is going on these water.

The water in oceans are vast, deep and uncontrolled stocking of fish seed will not profitable. Manuring of water cannot be done. Feeding of the required fish is impossible. Hence fish culture in oceans is not possible. However the culture of other aquatic organism i.e bivalue culture, Pearl culture in cages.

Based on Stocking Organism

The following aquaculture practice are followed:

- Monoculture: Rearing of a single species of fish is called Monoculture, it is called monospecies culture, e.g. Culture of Catla in a pond.

- Monosex culture: Rearing of only sex of species is called Monosex culture e.g. Tilapia.
- Polyculture: Rearing of different species in a water body is called polyculture. It is also called composite culture, e.g. culturing of catla, Rohu, Mrigal in a water body.
- Integrated fish culture Culturing: Culturing fishes along with paddy, poultry, piggery and dairy is called integrated fish culture, e.g. Indian major carps.

  (a) Paddy cum fish culture: Rearing of fish in paddy fields is called paddy cum fish culture, e.g. Catla, Rohu.

  (b) Poultry cum fish culture: Rearing of fishes along with poultry is called poultry cum fish culture, e.g. India major carps.

  (c) Dairy cum fish culture: Rearing of fishes along with dairy farm is called dairy cum fish culture, e.g. Indian major carps.

  (d) Pig cums fish culture: Rearing of fishes along with pig farm is called pi cum fish culture, e.g. Indian major carps.

Selection Criteria of Cultivable Species

Profitably culture species should have the following characteristics:

- Should be able to linen captivity (ponds, bunds, reservoir etc) with other fishes without any disturbance.
- Should be able to feed on natural as well as artificial. They should also be able to consume small quantity.
- Should be able to grow at faster rate and attain marketable is size shorter spam of time.
- Should be able to breed successfully and prolifically in confinement.
- Should be hardy and able to total rate climate as well as environment or ecological changes in the cultivable waters.
- Should be resistant to disease.
- Should be palatable and nutritive.

Aquaculture Methods

A number of aquaculture practices are used world-wide in three types of environment (freshwater, brackishwater and marine) for a great variety of culture organisms. Freshwater aquaculture is carried out either in fish ponds, fish pens, fish cages or, on a limited scale, in rice paddies. Brackishwater aquaculture is done mainly in fish ponds located in coastal areas. Marine culture employs either fish cages or substrates for molluscs and seaweeds such as stakes, ropes and rafts.

Culture systems range from extensive to intensive depending on the stocking density of the culture organisms, the level of inputs and the degree of management. In countries where government priority is directed toward increased fish production from aquaculture to help meet domestic demand, either as a result of the lack of access to large waterbodies (e.g., Nepal, Central African Republic) or the over-exploitation of marine or inland fisheries (e.g., Thailand, Zambia), aquaculture practices are almost exclusively oriented toward production for domestic consumption.

These practices include:

1. Freshwater pond culture;
2. Rice-fish culture or integrated fish farming;
3. Brackishwater finfish culture;
4. Mariculture involving extensive culture and producing fish/shellfish (e.g., oysters, mussels, cockles) which are sold in rural and urban markets at relatively low prices.

Aquaculture production systems and practices, by region.

Region	Major Culture Species	Major Culture Systems	Major Culture Practices	Scope for Future Development/Needs for Further Expansion
Asia	At least 75 species; diverse freshwater and marine species, including high-value shrimps, molluscs, seaweeds, with carps and seaweeds dominating production	Traditional extensive to intensive	- Fish ponds - Fish pens and fish cages - Floating rafts, lines, and stakes for molluscs and seaweeds	Development of culture-based fisheries in inland lakes, rivers, floodplains, and permanent and temporary reservoirs and barrages
				Resource enhancement programmes integrated with environmental management
Pacific	Mussels and oysters, red seaweeds	Intensive/ semi-intensive to extensive	- Hanging lines for mussels and pearl oysters	Production of high-value species for select markets;
			- Offshore cages for salmon	Small-scale aquaculture for local markets;
			- Pond culture for shrimps, tilapia, catfish, milkfish	Improved management of fishery resources, particularly reef fisheries
			- Freshwater pens for crayfish	
Latin America	50 species of fish, crustaceans, and molluscs, including freshwater fish and marine shrimps in South America and molluscs in Central America	Extensive to semi-intensive and Intensive	- Offshore cage farming of Pacific and Atlantic salmon - Ocean ranching in Southern Ocean - Semi-intensive farming of marine shrimp in coastal ponds and extensive farming of freshwater fish in ponds	Production of species for export and marine shrimp and salmon

Africa	>26 freshwater fish; the most important being tilapia and common carp, molluscs and oysters also	Mainly extensive, rural-based, integrated with poultry and animal husbandry, rice-fish farming; some intensive in raceways and floating cages	- Fish pond culture for freshwater fish - Raceways and floating cages for marine species	Increased emphasis on higher value catfishes for urban markets, on marine species of fish and crustaceans for select national market and export Culture-based fisheries in lakes and reservoirs Development of coastal lagoons which are almost totally unexploited
Mediterranean	>50 individual species, mostly freshwater and brackishwater fishes - most important being salmonids and carps; oysters and mussels	Well-diversified modern practices, with highly technical and intensive systems in developing countries and semi-intensive and extensive elsewhere	- Fish pond - Fish cages - Ocean ranching	Production of high-value species of tourism and export Integrated coastal zone management
Caribbean	About 16 species of tilapias, carps, marine shrimp and, freshwater prawns, oysters and seaweeds		- Floating cages in reservoirs	Priority is for aquaculture production for local markets
			- Fish pond farming in freshwater	
			- Culture-based fisheries in reservoirs	
			- Rope production of molluscs	

Extensive systems use low stocking densities and no supplemental feeding, although fertilization may be done to stimulate the growth and production of natural food in the water. Water change is effected through tidal means, i.e., new water is let in only during high tide and the pond can be drained only at low tide. The ponds used for extensive culture are usually large (more than two ha) and may be shallow and not fully cleared of tree stumps. Production is generally low at less than 1 t/ha/y.

Semi-intensive systems use densities higher than extensive systems and use supplementary feeding. Intensive culture uses very high densities of culture organism and is totally dependent on artificial, formulated feeds. Both systems use small pond compartments of up to one ha in size for ease of management.

Semi-intensive and intensive culture systems are managed by the application of inputs (mainly feeds, fertilizers, lime, and pesticides) and the manipulation of the environment primarily by way of water management through the use of pumps and aerators. Feeding of the stock is done at regular intervals during the day. In intensive shrimp culture, the computed daily feed ration is given in equal doses from as low as three to as high as six times a day. Water change is also effected on a daily basis, with approximately 10-15% of the water in the pond replenished by the entry of new water in semi-intensive shrimp ponds.

Semi-intensive and intensive culture systems are therefore more labour-intensive than extensive systems which need little attention, and are costlier to set up and operate, not to mention the fact that they also carry higher risks of mortalities resulting from disease, poor management, and/or force majeure (e.g., from anoxia due to non-functioning aerators during times of power failure).

Production is of course much higher (for example, ranging from a minimum of 1.5 t/ha/crop from semi-intensive shrimp culture to a high of 10 t/ha/crop from intensive shrimp culture). Financial returns are therefore much more attractive than those from extensive culture, although studies have shown that the return on investment (ROI) from semi-intensive culture is better than from intensive culture due to the high cost of inputs (largely fry and feeds) used in intensive culture.

Fish Pond Culture

Pond culture, or the breeding and rearing of fish in natural or artificial basins, is the earliest form of aquaculture with its origins dating back to the era of the Yin Dynasty. Over the years, the practice has spread to almost all parts of the world and is used for a wide variety of culture organisms in freshwater, brackishwater, and marine environments. It is carried out mostly using stagnant waters but can also be used in running waters especially in highland sites with flowing water.

Comparative features among the three main culture systems.

Parameter	Extensive	Semi-Intensive	Intensive
Species Used	Monoculture or Polyculture	Monoculture	Monoculture
Stocking Rate	Moderate	Higher than extensive culture	Maximum
Engineering Design and Layout	May or may not be well laid-out	With provisions for effective water management	Very well engineered system with pumps and aerators to control water quality and quantity
	Very big ponds	Manageable-sized units (up to 2 ha each)	Small ponds, usually 0.5-1 ha each
	Ponds may or may not be fully cleaned	Fully cleaned ponds	Fully cleaned ponds
Fertilizer	Used to enhance natural productivity	Used regularly with lime	Not used
Pesticides	Not used	Used regularly for prohylaxis	Used regularly for prophylaxis
Food and Feeding Regimen	None	Regular feeding of high quality feeds	Full feeding of high-quality feeds
		Depending on stocking density used, formulated feeds may be used partially or totally	
Cropping Frequency (crops/y)	2	2.5	2.5
Quality of Product	Good quality	Good quality	Good quality
	Culture species dominant but extraneous species may occur	Confined to culture species	Confined to culture species
	Variable sizes	Uniform sizes	Uniform sizes

Running water fish culture involves growing the fingerlings to marketable size in earthen ponds using water from rivers, irrigation canals, or plain rain water. The system approximates intensive culture in that it involves the application of rapid water changes and the heavy stocking of the cultured species. The continuously flowing water is advantageous for fish culture as it supplies abundant dissolved oxygen and flushes away waste products and unconsumed feeds.

Culture Species

Commonly raised species in freshwater ponds are the carps, tilapia, catfish, snakehead, eel, trout, goldfish, gouramy, trout, pike, tench, salmonids, palaemonids, and the giant freshwater prawn Macrobrachium. In brackishwater ponds, common species include milkfish (Chanos chanos), mullet (Mugil sp.) and the different penaeid shrimps (Penaeus monodon, P. orientalis, P. merguiensis, P. penicillatus, P. semisulcatus, P. japonicus, and M. ensis). The more popular species for culture in marine ponds are the sea bass, grouper, red sea bream, yellowtail, rabbitfish, and marine shrimps.

In Asia, where the bulk of world production from aquaculture emanates, fish ponds are mostly freshwater or brackishwater, and rarely marine. In China and most of the Indian sub-continent, pond culture is traditionally dominated by freshwater species, mainly the carps, usually in polyculture and/or integrated with animal husbandry. In Southeast Asia, fish ponds are predominantly brackishwater, with milkfish and penaeid shrimps grown either in polyculture or in monoculture.

Recently in Latin America and the Caribbean, brackishwater pond culture of penaeid shrimps has expanded rapidly, as it has in some parts of Asia.

In Africa, the tilapias and carps dominate aquaculture production. Controlled breeding is also carried out in ponds with goldfish, trout, Bagrus and, to a lesser extent, Lates niloticus, Heterotis niloticus, and Clarias lazera. Ten species of molluscs belonging to four genera (Crassostrea, Mytilus, Venerupis and Pinctada) are cultured. Crustacean culture has yet to be developed on a significant scale.

Site Selection

Proper site selection is recognized as the first step guaranteeing the eventual success of any aquaculture project and forms the basis for the design, layout, and management of the project. For fish ponds, especially those to be used for coastal/brackishwater aquaculture of high-value species like shrimps, site selection is critical and should be given utmost attention.

Following are the guidelines for the selection of a suitable site for coastal fish ponds:

1. Soil Quality: Preferably, clay-loam, or sandy-clay for water retention and suitability for diking; alkaline pH (7 and above) to prevent problems that result from acid-sulphate soils (e.g., poor fertilizer response; low natural food production and slow growth of culture species; probable fish kills).
2. Land elevation and tidal characteristics: Preferably with average elevation that can be watered by ordinary high tides and drained by ordinary low tides; tidal fluctuation preferably moderate at 2-3 m. (Sites where tidal fluctuation is large, say 4 m, are not suitable because they would require very large, expensive dikes to prevent flooding during high tide. On the other hand, areas with slight tidal fluctuation, say 1 m or less, could not be drained or filled properly).
3. Vegetation: Preferably without big tree stumps and thick vegetation which entail large expense for clearing; areas near river banks and those at coastal shores exposed to wave action require a buffer zone with substantial growths of mangrove. (The presence of Avicennia indicates productive soil; nipa and trees with high tannin content indicate low pH).

4. Water supply and quality: With steady supply of both fresh and brackish water in adequate quantities throughout the year; water supply should be pollution-free and with a pH of 7.8-8.5.

5. Accessibility: Preferably readily accessible by land/water transport; close to sources of inputs such as fry, feeds, fertilizers, and markets, fish ports, processing plants, and ice plants; and linked by communication facilities to major centres.

6. Availability of manpower for construction and operation.

Pond Layout

The layout of the pond system depends on the species for culture and on the size and shape of the area, which in turn determines the number and sizes of ponds and the position of the water canals and gates. A fish farm is considered properly planned if all the water control structures, canals, and the different pond compartments mutually complement each other. A complete fish farm has nursery and grow-out ponds and, in some instances, transition ponds for intermediate-sized fish/shrimp, all of which are properly proportioned and positioned within.

Milkfish culture in brackishwater ponds in the Philippines follows the traditional practice of providing for nursery, transition, and rearing operations. In some cases, formation ponds are used for additional growth or stunting of fingerlings prior to stocking in rearing ponds. The nursery ponds comprise about 1-4% of the total production area while the transition and formation ponds constitute about 6-9% of total area.

It has been suggested that a similar progressive culture scheme be adopted for shrimp pond culture when no supplementary feeding is practised. For growing to a medium size, a two-stage progression composed of a nursery pond (NP) and a rearing pond (RP) is adequate; for growing to larger sizes, a three-stage progression composed of nursery, transition, and rearing ponds is recommended.

In general, however, shrimp monoculture uses direct stocking of post larvae in rearing ponds and therefore requires only one type of pond with separate inlets and outlets for better circulation and aeration.

Design of Pond Facilities

A fish pond system consists of the following basic components:

1. Pond compartments enclosed by dikes;

2. Canals for supply and drainage of water to and from the pond compartments; and

3. Gates or water control structures to regulate entry and exit of water into and from the pond compartments.

Pond compartments are usually rectangular in shape although in Indonesia, running water ponds are generally triangular, raceway-shaped, or oval. They vary in size from less than a hectare to several hectares each, sometimes up to 20-50 ha in size. However, with the new intensive methods, the trend is to use smaller units for flexibility and ease of management.

The elevation of the rearing pond bottom for milkfish is usually such that only a maximum of 40 cm of water can be held in the ponds during the culture period. For new shrimp ponds, the minimum water depth is 1 m.

The entire pond system is enclosed by a perimeter dike and the individual pond compartments are separated from each other by partition dikes. The outer perimeter dike is usually wider and higher than the inner partition dikes and serves to protect the entire fish pond area from flooding and destruction brought about by tide and wave action. The inner dikes are narrower and shorter.

The design of the dikes depends primarily on soil characteristics. Dikes are usually earthen although intensive shrimp ponds are concrete-lined or brick-lined as in Taiwan (PC). The side slopes are designed for structural stability, the ratio of horizontal length to height ranging from 1:1 to 1:3. The height and width of dikes depend on the type (primary, secondary, or tertiary), tide conditions, flood level, pond water depth, soil shrinkage, and freeboard.

The following slopes are recommended for dikes built with good clay soil:

- 2:1 when dike height is above 4.26 m and exposed to wave action;
- 1:1 when dike height is less than 4.26 m and tidal range is greater than 1 m; and
- 1:2 when tidal range is 1 m or less, and dike height is less than 1m.

The dike crown should not be less than 0.5 m and the main dike surrounding the farm should be 0.5 m above the highest dike or flood level recorded in the locality.

Water conveyance structures (canals/channels) supply new water into the pond and drain out old water. They also provide the facility for holding and harvesting of fish and of serving as waterways for transporting farm supplies. Traditional milkfish ponds usually have only one canal that is used for both supply and drainage. Shrimp ponds have separate supply and drainage canals. Canals which are to be used for harvesting should be 30 cm below the level of the pond bottom to allow draining of pond water.

Having separate water intake and discharge canals in a pond complex brings about the following advantages:

- Better filling and non-contamination of pond by discharge from other ponds.
- Greatly reduced possibility of spread of disease.
- Maintenance of constant head in intake canal thus reducing water loss through leaks/seepages in pond dikes and consequently reducing leaching of acids into the ponds from dikes with acid-sulphate soils.
- Absence of conflict of usage between farmers.
- Better water exchange for individual ponds, and
- Possibility of effecting flow-through systems.

The width of the canals depends on the amount of water they must carry. The following should be taken into account when designing canals:

1. Volume of water to be held in the ponds.
2. Time requirement for filling or draining the pond.
3. Amount of rainfall which must be carried off in a given period of time.
4. Elevation of canal bottom in relation to tide.
5. Other uses like transportation, harvesting of milkfish, and holding of broodstock.

Diversion canals are constructed where there is much runoff from adjoining areas, to prevent sudden salinity changes and the possible entry of polluted, pesticide-loaded water and/or of silted water into the pond complex.

The entry and exit of water into ponds through the canals is regulated or controlled by gates. Main gates regulate the exchange of water between the pond system and the tidal stream or sea, and may be constructed of reinforced concrete or wood. Reinforced concrete is more expensive but lasts longer. Such a gate has one or multiple openings depending on the relative size of the pond unit to be served. A recent innovation for a smaller and less expensive main gate is the monk-type gate which uses culverts usually made of concrete hollow blocks. The SEAFDEC Aquaculture Department has also introduced the open sluice gate made of ferrocement.

Secondary gates, which regulate water exchange between the ponds and the canals, are usually made of wood. Pipes or culverts can also be used for smaller ponds such as nursery or fry ponds and transition ponds for milkfish culture. Secondary gates are now usually located toward one end of the narrower side of the pond compartment to give good turbulence and circulation during the filling and draining.

Shrimp ponds are provided with separate supply and drainage gates to effect flow-through water management and facilitate water exchange through supply and drainage canals. Inlet and outlet gates are best located at opposite corners of the same pond, across which a diagonal trench, about 5-10 m wide and 0.3-0.5 m deep, extending from inlet to outlet gates is recommended for convenient draining of water.

Gates should be located where they are not exposed to strong weather forces and where water of good quality can be allowed to enter the fish pond system. Proper gate location can also serve to aerate the pond water and promote water circulation.

During the construction of gates for shrimp ponds a number of requirements should be kept in mind and the gates should:

1. Be durable, water-tight, and made of locally available materials;
2. Have adequate capacity for the amount of water to be taken in or drained;
3. Allow water to be taken in or discharged at the bottom;
4. Have provisions for draining pond surface water;
5. Have gate bottom elevation that permits complete draining of pond water;
6. Have slots or grooves for the placement of outside and inside screens to prevent undesirable species from entering the pond and the shrimps from leaving the pond;
7. Have place for net installation for harvesting; and
8. Be easy to operate.

Pond Management

Pond management techniques for finfish and shrimp culture, while varying slightly depending on the specific biological requirements of the culture organism, the type of culture system, and the culture environment (freshwater, brackishwater, and marine), are similar in that they involve the following basic activities:

1. Pond preparation/conditioning.
2. Stocking.
3. Feeding and/or fertilization (depending on the culture system used).
4. Water management.
5. Pond maintenance, and
6. Harvesting.

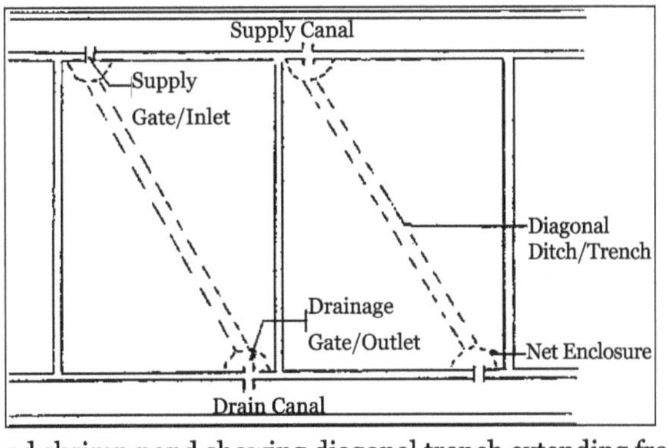

Layout of improved shrimp pond showing diagonal trench extending from inlet to outlet

Variations would consist mainly of differences in application rates of fertilizers, lime, pesticides, and feeds; stocking rates and sizes of stocking material; rate of water change; and harvesting techniques.

In extensively managed systems generally require the least management, with no supplemental feeding and minimal water exchange on account of the low stocking density used. On the other hand, intensively managed ponds require full artificial feeding and substantial water management to ensure optimum culture conditions for the species being reared.

Pond Preparation

Ponds are totally drained and the pond bottoms dried prior to the application of pesticides. Tobacco dust, derris root/rotenone powder, teaseed cake/powder, or Gusathion-A are used to eliminate predators and/or wild species that may eventually compete with the cultured organisms for food and space. Teaseed cake is perhaps the best fish poison to use in brackishwater ponds to selectively kill unwanted fish without damaging the shrimps and without affecting rotifers and copepods which are feed for shrimps. On the other hand, rotenone is most effective in fresh water and works better in low-salinity water.

Ponds with acid-sulphate soils are repeatedly dried and flushed, i.e., filled and drained to remove the acids formed by pyrite oxidation. Agricultural lime is then applied to correct soil pH and bring it up to at least 6.5. Brackishwater ponds are usually treated by spreading 1.5 t of agricultural lime per ha, followed by another 1.5 t worked into the soil.

To stimulate and maintain the growth of natural plankton, organic (e.g., chicken manure) or inorganic fertilizer (e.g., urea, ammonium phosphate) are applied to the pond bottom. After fertilizer application, water is let in to a depth of about 20-40 cm and gradually increased to 1 m a week after fertilization. Intensively managed ponds or ponds where artificial feeding shall be given, do not need to be fertilized. Extensive ponds need regular fertilization during the culture period to maintain the growth of natural food. Semi-intensive ponds may use a mix of fertilization and supplementary feeding.

Variations in pond management techniques commonly used for different species.

Species	Stocking Rate	Fertilization	Feed Type	Rate of Water Change	Pesticides/Predator Control	
					Type	Application Rate
Milkfish (Chanos chanos)	2 000-5 000/ha	16-20-0 at 50 kg/ha; 45-0-0 at 15 kg/ha; chicken manure at 0.5 t/ha twice weekly	Rice bran and trash fish as supplemental feed	Once every two weeks at high tide	Lime; ammonium sulfate	1 t/ha 10 g/m^2
Tilapia (O. niloticus; O. mossambicus)	5 000-20 000/ha	Chicken manure at 500 kg/ha; Inorganic fertilizers at 50 kg/ha	Rice bran, fish meal, ipil-ipil leaf meal			
Catfish (Clarias-botrachus and monocephalus)	60-300/m^2		9 parts trash fish and 1 part rice by-products	When necessary		

Penaeids	From as low as 15 000 to as high as 300 000/ha	Chicken manure at 1-2 t/ha followed by inorganic fertilizer at 75-150 kg/ha mono-ammonium phosphate (16-20-0) and 25-50 kg/ha of urea (46-0-0)	Supplemental feed of rice bran with trash fish, mussels, and clam meat; artificial/formulated diets with 40% CP.	20-30% once every week or every two weeks for low density ponds; 5-20% daily for semi-intensive to intensive ponds		

Stocking

After the pond is prepared, fish fingerlings or shrimp post larvae are stocked at the appropriate density depending on the culture strategy, size of pond, and the size of fingerlings, among others.

The fingerlings are properly acclimated and conditioned prior to stocking and weak or diseased fish eliminated. Stocking is usually done in the early morning or late afternoon.

Feeding

Fish/shrimp grown in semi-intensive and intensive culture ponds are given supplementary and full artificial feeds, respectively, the former to augment the natural food in the pond, the latter to totally replace the natural organisms in the water as a source of nutrition.

A wide variety of feed ingredients is used to prepare supplemental/artificial feeds. The simplest fish feeds are prepared at the pond site using locally available raw materials like rice or corn bran, copra meal, and rice mill sweepings as sources of carbohydrates. These are usually mixed with animal protein like trash fish/fish meal, shrimp heads, and snail meat. Supplemental feeds for tilapia are prepared using 80% rice bran and 20% fish meal. Those for shrimps in improved extensive culture (low-density stocking but given dietary supplements for increased growth/production) usually include fresh raw materials like snail/mussel/clam meat or carabao hide and other slaughterhouse leftovers.

Commercial feed preparations are also available now in a wide range of brandnames, mostly for semi-intensive and intensive shrimp culture. (Taiwan (PC), Japan, and the USA are the top producers of commercial fish/shrimp feeds). These commercial diets consist of a number of ingredients like fish meal, blood meal, bone meat, and shrimp head meal (to serve as attractant for the shrimp), together with vitamin and mineral premix and carbohydrate sources like rice/corn bran or wheat. The crude protein (CP) content of these shrimp feeds is generally not lower than 30% to satisfy the high animal protein requirement of shrimps, actually estimated to be about 40% during the earlier stages of growth.

Commercial feeds usually come in various formulations to match the protein requirement of the culture organism, which as a rule, decreases with age. Thus, fish/shrimp feeds come in different forms as starter, grower, and finisher, with starter feeds having the highest CP content of about 40% and finisher feeds having the lowest CP content of about 20%. Starter feeds are usually given on the first month of culture, finisher feeds on the last month, and grower feeds in between.

Some shrimp culturists prefer not to give artificial feeds during the first two weeks of culture when the newly stocked post larvae can subsist on the plankton available in the water.

The feeding rate is computed as a percentage of the estimated animal biomass in the pond, with higher rations given when the animals are small and gradually decreasing as they become bigger. The daily feeding rate usually starts at 5% and 10-15% of estimated biomass of fish and shrimps, respectively, and decreases to a low of 2% and 5%, for fish and shrimps, respectively, toward harvest.

The daily feed rations are given in equal portions during the course of a day. Freshwater fish like tilapia are usually fed twice a day - early morning and late afternoon. Penaeid shrimps are fed more frequently, from three to four to as often as six to seven times a day.

Feeds are broadcast into the water and/or supplied on feeding trays. In semi-intensive and intensive shrimp ponds, small feeding boats are used by caretakers who go around the pond distributing the feed by broadcasting. At certain points along the periphery of the pond, feeding trays are submerged into the water after known quantities of feed are put on the surface, to supply feed to the shrimps in the pond as well as to monitor feed consumption and shrimp growth. The feeding tray is lifted two to three hours after the feed was supplied to check how much of it has been consumed and to see if the shrimps are healthy and feeding. Empty feeding trays may indicate that the quantity given is inadequate and may have to be increased. Conversely, full or slightly touched trays indicate excessive feed quantities and/or sluggish shrimps. The feeding ration is subsequently adjusted accordingly to optimize feed utilization.

By monitoring the feeding tray, one can get a good indication of the sizes and quantity of shrimps present in the pond without a need for cast-netting or actual sampling, since shrimps are invariably found on the tray when it is lifted out of the water.

Water Management

Water in the pond is kept at certain levels for optimal fish growth. In general, a pond water depth of 1 meter is considered best for culture of tilapia, carps, and shrimps; traditional milkfish ponds can do with just 40-60 cm of water.

Pond water is not just maintained at a certain depth; its quality must also be kept high to ensure optimal growth of the culture organism. This is particularly important in semi-intensive and intensive culture systems where large amounts of metabolites are continously excreted into the pond and where excess, unconsumed feeds add to the bottom load and serve to pollute the water.

To prevent the deterioration of the pond environment, pond water is continuously freshened by the entry of new water from the river or water source (through the supply canal) while old water is drained through the outlet/drainage gate and through the drainage canal into the sea or river.

A flow-through system of water management that allows the simultaneous entry and exit of water into and out of the pond is essential in any high-density culture system. This is effected by the provision of separate inlets and outlets for all the ponds, each inlet regulating the flow of water from the supply canal to the pond and each outlet controlling the discharge of water out of the pond into the drainage canal. Both the supply and drain gates are so designed as to bring water into and drain water out of the lower levels of the pond, where water quality tends to get poorer faster as a result of the accumulation of wastes and their subsequent decomposition.

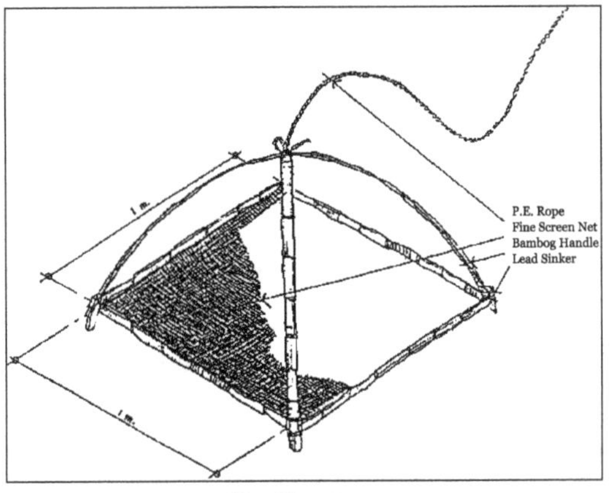

Feeding tray

The regular replenishment of pond water, independent of natural tidal fluctuations, is made possible by the use of pumps which draw water from the source even at low tide. Although there is no hard-and-fast rule as to the rate of water change necessary for medium- to high density aquaculture, semi-intensive culture systems usually change water at the rate of 10% daily for an equivalent total replacement of water every ten days or three times per month. Intensively managed ponds require greater water exchange in view of the much higher organic load on the pond bottom, especially toward the latter part of the culture cycle when the animals excrete more wastes.

Intensive ponds/tanks usually need to provide for aeration facilities/equipment to prevent anoxia that may lead to mass mortalities. Oxygen depletion in high-density ponds results not only from the faster rate of utilization of dissolved oxygen for respiratory activities; it is also caused by the fast rate of decomposition at the pond bottom by aerobic or oxygen-consuming micro-organisms.

Paddlewheels or other types of aerators are thus provided in the ponds to effect the infusion/introduction of greater quantities of oxygen into the water and prevent fish/shrimp mortalities. The aerators are usually operated at regular/periodic intervals for certain fixed durations during the day but especially in the early morning hours when the concentration of dissolved oxygen is known to be lowest (as a result of the absence of photosynthetic, oxygen-producing activity in the pond). Toward the end of the culture period when oxygen demand is highest, aeration may have to be provided continuously and not just sporadically as could be done during the initial stages of rearing. At that time too, water pumps usually need to be run for longer periods to effect greater water exchange.

Pond water is also regularly sampled and measurements taken of basic/essential parameters, particularly dissolved oxygen, pH, and salinity. This is important for the purpose of determining the need for corrective/remedial action to bring water quality to optimum levels and obtain good yields.

Dissolved oxygen levels are kept, as much as possible, above 5 ppm by pumping and aeration. Problems of acidity are corrected by liming. Salinity is an important parameter for penaeid culture and has to be maintained within a range of 15-25 ppt for best results. During summer months, high-salinity water can be diluted by mixing with fresh water from springs or deep wells.

Pond Maintenance

Fertilization

Aside from feeds and water management, the following pond maintenance procedures are carried out: regular application of fertilizers, lime, and pesticides; prevention of entry of predators; monitoring of the stock for growth rate determination as a basis of feeds and water management; and regular pond upkeep and maintenance.

Extensive ponds are fertilized regularly using either organic fertilizers like chicken, cow, or pig manure, or inorganic fertilizers like urea, ammonium phosphate, or both, to maintain the plankton population in the pond. The fertilizers are either broadcast over the pond water surface or kept in sacks suspended from poles staked at certain portions along the pond periphery. Semi-intensive and intensive culture systems do not require fertilization since they are not natural food-based, except for those which grow plankton-feeders like milkfish whose diet is largely algae dependent.

Liming

In addition to fertilization, ponds also need to be given regular doses of lime to maintain water pH at alkaline or near-alkaline levels (preferably not lower than six). Agricultural lime is broadcast over the pond and applied on the sides of the dikes to correct soil and water acidity.

Elimination of Pests and Predators

Unwanted and predatory species which may have survived the application of pesticides during pond preparation or which were able to enter the pond through the gate screens or through cracks in the dikes, are eliminated by the application of pesticides, preferably organic, into the pond.

Crabs, which are a serious problem in shrimp ponds because they are carnivorous and cause damage to the pond dikes, are not usually affected by known pesticides and are therefore best eliminated by the use of crab traps situated in the pond.

It is also important that the gates are properly screened and the screens kept whole, to prevent the entry of small unwanted fish into the pond. Double screens are usually installed at the main intake to ensure that pests and predators are prevented from entering the pond system.

Stock Monitoring

The culture organisms are monitored closely and regularly to determine their rate of growth and the general condition of the stock. They are regularly sampled for length-weight measurements as a basis for determining/estimating their biomass in the pond and therefore their daily feed rations, as well as for making projections on harvest schedules and procurement of pond inputs.

In the first few months of culture, the feeding tray is a good tool for stock monitoring. As the organisms grow in size, cast-netting is used as a sampling tool, with those caught in the throw of the cast net providing an indication as to sizes and weights of stock. Based on the sampled weights and the daily feed consumption, it is possible to predict the available biomass (i.e., stock surviving after initial mortalities) and make projections on volume of harvest. For this purpose, it is essential that accurate records are kept for analysis at a later time. Data on initial size/weight and number of fry/post larvae stocked, average body weight at each sampling, and feed consumption on a daily basis, are important to have on file.

Regular Upkeep and Maintenance of Facilities

The pond dike and gates are checked regularly for cracks that could lead to seepages and losses of stock. The dikes are best planted with grass or vegetative cover to prevent erosion. The gates and other support infrastructure are properly maintained for efficient operation.

Harvesting

Marketable-size fish/shrimps are harvested at the end of the culture period by draining the pond and using harvesting nets to catch the fish or shrimps. The latter are harvested with a bag-net attached to the sluice gate as water is drained out of the pond at low tide. Tilapia are harvested using seine nets after the pond water is drained to half-level the night before.

Harvest of milkfish takes advantage of their behaviour of swimming against the current. The method, known in the Philippines as "pasulang" or "pasubang" involves draining 85-90% of the pond water during low tide and allowing in the water at the incoming high tide so that the fish swim against the current through the tertiary gate and into the catching pond, whose gate is closed once a large number of fish is impounded. The fish in the catching pond are then harvested by seining and the rest hand-picked.

Integrated Fish Farming

In a number of countries in Asia (e.g., China, Nepal, Thailand, Malaysia, Indonesia) and in some parts of Africa, freshwater fish culture is integrated with the farming of crops, mainly rice, vegetables and animals (usually pigs, ducks, and chickens). This leads to greater overall efficiency of the farming system as wastes/by-products or one component are used as inputs in another. For example, poultry or pig manure can be used to fertilize the fish pond and the vegetable garden and the waste vegetables can be fed to the fish and the pigs.

In Africa, fish culture in rice fields and in combination with pig and duck rearing, is not too widely practised but has significant potential. Reported fish yields ranged from 2 000-4 000 kg/ha/y with ducks, 8 500-8 900 kg/ha/y with pigs, and 3 600-4 900 kg/ha/y with poultry in Gabon. It has also been proven economically viable since it involves minimal investment. Its spread has, however, been constrained by the widespread use of pesticides in many countries.

Pen and Cage Culture

Pen and cage culture involve the rearing of fish within fixed or floating net enclosures supported by frameworks made of bamboo, wood, or metal, and set in sheltered, shallow portions of lakes, bays, rivers, and estuaries.

Compared to fish pond culture with its 4 000-year tradition, fish pen/cage culture is of more recent origin. Cage culture seems to have developed independently in at least two countries - in Kampuchea where fishermen in and around the Great Lake region would keep Clarias spp. and other commercial fishes in bamboo or rattan cages and baskets; and in Indonesia where bamboo cages have been used to grow Leptobarbus hoeveni fry as early as 1922. Since then, cage culture has spread throughout the world to more than 35 countries in Europe, Asia, Africa, and the Americas.

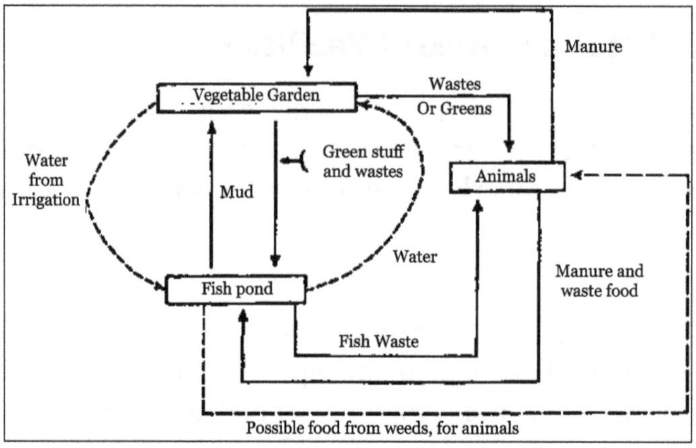

Diagram showing interrelationships among the various components of an integrated fishfarming system.

Pen culture is said to have originated in the Inland Sea area of Japan in the early 1920s, adopted by the People's Republic of China in the 1950s for rearing carps in freshwater lakes, and introduced to culture milkfish in the shallow, freshwater, eutrophic Laguna de Bay in the Philippines in the 1970s. From there it has been successfully extended for the culture of tilapia and carps. Its development and adoption as a popular technology has not been widespread, though, perhaps because of its site-specific requirements like its suitability mainly in shallow lentic environments. At present, it is commercially practised only in the Philippines, Indonesia, and China.

The wider popularity of cage culture as compared to pen culture may be due to its greater flexibility in terms of siting the structures. For example, cages may be installed in bays, lagoons, straits, and open coasts as long as they are protected from strong monsoonal winds and rough seas. Floating cages can also be set up in deep lakes and reservoirs, and in rivers and canal systems, and even in deep mining pools which could not be used otherwise for culture due to harvesting difficulties.

In general, however, both pen and cage culture have expanded rapidly, especially over the past two decades vis-a-vis the decreasing availability of land-based resources for fish culture and an increasing awareness of their merits over traditional pond culture, such as:

1. Their applicability in different types of open water bodies like coastal waters, protected coves and bays, lakes, rivers, and reservoirs;
2. Their high productivity (of as much as 10-20 times that of ponds of comparative sizes) with minimal inputs and at lower costs to develop and operate; and
3. The greater socio-economic opportunities they provide to low-income families in the rural areas, particularly those displaced by the reduction of fish catches in over-exploited coastal, municipal waters, because they require comparatively low capital outlay and use simple technology.

Yields from pen and cage culture are generally high, with or without supplemental feeding depending on the natural productivity of the water body. In the Philippines, for example, the yields of milkfish from fish pens in Laguna de Bay were as high as 4 t/ha/crop (compared to a national milkfish fish pond average of 1 t/ha/y in 1980 when the productivity of the lake was very high at 1 700 mg C/m³/hr.

In Indonesia, the cage culture of common carp in the Lido Reservoir in Cigombong gave a total production of 28 kg/m² at a stocking density of 6 kg/m². The cage culture of marine finfishes has likewise been shown to give high yields.

Culture Species

The choice of species for stocking and rearing in pens and cages is governed by much the same criteria as in species selection for pond culture, including:

1. Fast growth in confinement;
2. Good consumer acceptance;
3. high tolerance to a wide range of environmental conditions;
4. Resistance to disease;
5. Ready supply of fish seed for stocking; and (vi) ease of culture and management.

Comparison of production of cage-cultured marine fish.

Species	Seriola T: quinquer-adiata	Trachinotus caro-linus	Polydactylus sexfilis	Epinephelus sal-moides*
Country of culture	Japan	Florida, USA	Hawaii, USA	Penang, Malaysia
Initial stocking density				
fish/m³	10	250	50	60
kg/m³	0.15-0.55	1.75	0.4	3.4
Rearing period (days)	225	273	300	240
Production (kg/m³)	0.85-14.45	44.7	-	41.4
Average production rate (kg/m³/day)	0.004-0.06	0.16	-	0.17
Mean size of fish				
Initial (g)	10-50	7	9	55.7
At harvest (g)	1 000-2 000	213.6	300	795.9
Average growth rate (g/fish/day)	4.40-8.67	0.76	0.97	3.08

There are approximately ten species of fish which are commercially cultured in cages and pens in both temperate and tropical waters, including tilapias (S. mossambicus and S. niloticus); carps (Chinese, Indian, and common varieties); milkfish; snakeheads and catfishes; marble goby; and salmonids (rainbow trout, salmon). Marine species include mainly grouper, sea bass, mullet, snapper, and milkfish.

In the Philippines, Indonesia, and China, pen culture is limited to the following species: milkfish (Chanos chanos); tilapia; and the Chinese carps: bighead (Aristichthys nobilis), silver carp (Hypophthalmichthys molitrix), grass carp (Ctenophanyngodon idella); and common carp (Cyprinus carpio).

Other species have been suggested as possible candidates for utilization in pen/cage culture in the following three different environments:

- Freshwater:

  Habitats with high natural productivity (e.g., lakes, oxbow lakes, swamps, mining pools, rivers, and reservoirs): mullets, eels, catfish, Puntius gonionotus.

  Habitats with low natural productivity: Leptobarbus, Clarias batrachus, Oxyeleotris, and Macrobrachium.

- Brackishwater:

  Sea bass, mullet, siganids, sea bream, grouper, snapper, threadfin, carangids. Hilsa spp., Sparus spp., and eels.

- Marine:

  Siganids, pampano, yellowtail, tuna, grouper, snapper, sea bass, sea bream, carangids, pomfret.

Site Selection

The selection of sites for fish pen/cage culture should be guided by the following basic criteria:

- Protection from high winds or typhoons.
- Adequate water exchange that will enable the flow of nutrient-laden water through the pens/cages.
- Good water quality (high or adequate dissolved oxygen, stable pH, and low turbidity, and absence of pollution).
- Firm bottom mud to allow pen framework to be driven deep into substrate for better support.
- Freedom from predators and natural hazards.
- Accessibility to sources of inputs, including labour and markets, and
- Good peace and order condition.

Design and Construction

Both fish pens and fish cages are built around the same basic design concept: a net enclosure supported by a rigid framework. They differ, however, in a number of respects. Firstly, a pen does not have a net bottom; the edges of its net wallings/fencings are anchored to the lake bottom/substrate by means of bamboo pegs and the lake bottom is the pen bottom. In comparison, a cage is like an inverted mosquito net with the cage bottom made of the same netting material used for its four sides.

Secondly, fish pens theoretically have no limit to their size/area while cages cannot exceed 1000 m^2 in area for reasons of the quantity of material required for cage construction (due to the need for a flooring) and manageability of operation (cages have to be lifted and the fish scooped out and not harvested using nets as in pens).

Thirdly, design of the structures and methods of construction are different. Fish pens are fixed structures; fish cages may either be fixed or floating. Fish pens for milkfish culture in Laguna de Bay, Philippines consist of a nursery pen within the grow-out pen/enclosure. Cages are individual units for either seed production or grow-out; they are, however, usually installed in clusters or modules with a common framework.

Pens and cages come in various shapes and sizes and are made of different types of materials. Most pens and cages are rectangular or square although some may be circular, as in some milkfish pens in Laguna de Bay and the milkfish broodstock cages at the SEAFDEC Aquaculture Department in the Philippines, or cylindrical as those used for fish collection in Malaysian or Indonesian fresh waters. Rectangular cages are preferred for easy operation and management. Circular cages are more suitable for some species like milkfish and yellowtail but are more expensive to build.

Polyethylene and nylon monofilament twine are widely used for fabricating cages and net pens although wire mesh is used in several countries. The framework structure is generally made out of bamboo and other locally available wood. Cage floatation materials include bamboo, PVC pipes/ containers, steel or plastic drums, styrofoam, and aluminum floats. The type of anchor for floating cages varies depending on the depth of water, nature of bottom, tides, and currents. Concrete slabs of different sizes and shapes, sand bags, and iron anchors are widely used in different countries.

Pen and Cage Operation

Basic procedures involved in the management of pen and cage culture are very much like those in pond culture, starting with completion of construction and preparation of the culture facilities for stocking, rearing, and harvesting. Slight variations in specific activities exist, however, as the result of the very nature of the system. For example, it is obviously not possible to apply fertilizers, lime, and pesticides since the system has open water exchange between the inner compartment and the outside environment.

Soon after construction of the pen/cage is completed, preparations are made to procure fry/fingerlings for stocking. Milkfish pens have a nursery compartment into which milkfish fry are grown for 3-4 weeks to 12 cm long fingerlings which can be released into the grow-out compartment.

The nursery pen and the grow-out compartment are prepared for stocking by clearing the bottom of predatory fish like Megalops cyprinoides and Elops hawaiiensis. The milkfish fry/fingerlings from the nursery pen are stocked in the rearing pen at 20,000-50,000 per ha where they are cultured to marketable size.

In the Philippines, the milkfish stock in the pen is not generally given supplemental feeding except for occasional rations of bread crumbs, rice bran, broken ice cream cones, fish meal, and ipil-ipil leaf mill.

On the other hand, cage-reared fish may or may not be fed supplemental or artificial diets depending on the stocking density used and the level of technology in the country. Cage feeding trials in the Philippines showed the adequacy of a ration composed of 77% rice bran and 23% fish meal with feed conversion ratios of 2.2-2.8. Current feed practices in freshwater cage culture involve the provision of supplemental feeds using readily available ingredients like rice bran and poultry feeds. Other countries use artificial feeds based on simple diets preferably prepared in pelleted form for best results.

At the end of the culture period, the fish are harvested from pens using harvesting nets (e.g., gill nets, cast nets, seines) or from cages by lifting the cage and causing the fish to collect in one corner for scooping out using a pail.

Feed types given to cage-reared fish.

Country	Culture Species	Feed Type
GDR	Common carp	Formulated feed/pellets, 33.7% CP
USSR	Common carp	Mixture of minced trash fish, molluscs, crayfish, and grown cereals
Hungary	Wels (Silurens glanis)	Trash fish, slaughterhouse wastes, cereal grain meals
-do-	Carp polyculture (common, silver, bighead)	Pelleted common carp feed
India	Indian carp polyculture	Soya bean powder, ground nut, oil cake, rice polish (1:1.1)
Indonesia	Leptobarbus hoeveni and Thynnichthys thynoides	Coconut water, cassava, rubber leaves
Indonesia	S. niloticus	Aquatic plants (Lemna, Hydrila, Chara)
Nepal	Common carp	Wheat flour, rice bran, mustard oil cake
Thailand	Catfish, sand goby, common carp, local carp, tilapia, snakehead	Pellets consisting of ground fish meal, soy bean, peanut, and rice bran
	Sea bass (Lates calcarifer)	Trash fish

Open Water Culture

The farming of molluscs and seaweeds in open marine waters has become increasingly popular in a number of countries, especially in the Third World where it is seen as a viable alternative to municipal or artisanal fisheries or as a means of supplementary income for small-scale fishermen. Because seafarming is generally low-cost and labour-intensive and could thus involve entire coastal communities, it is particularly appropriate in areas where production from municipal fisheries has substantially declined and where, as a result, subsistence fishermen have little or no means of livelihood.

Mollusc Culture

Bivalves are widely cultured in a number of countries world-wide. In Asia and the Pacific, they represent a high quality food resource with annual production higher than from crustacean culture on a per hectare basis. In 1984, molluscs accounted for approximately 35% of the total production of coastal aquaculture in terms of gross weight in the region.

The most important species for culture in Southeast Asia are the oysters (mainly Crassostrea spp.), mussels (mainly Perna spp.), clams, cockles, and scallops.

In Japan, the most commonly cultured species include Crassostrea gigas, C. rivularis, C. nippona, C. echinata, and Ostrea denseramellosa, with C. gigas as the predominant species. In Africa, the culture of Venerupis is reported in Tunisia and Pinctada spp. in Sudan. In Mexico, the culture of the large oyster Crassostrea spp. is carried out by cooperative societies and of the mussel Mytilus edulis on floating rafts by private investors.

Oysters are widely distributed in estuaries and bays which receive some run-off from land and have somewhat lower salinity than the open sea. As they filter their food from the water, they grow best in areas with moderate to high concentrations of phytoplankton. Oysters grow best in inter-tidal areas where they are exposed for some minutes or a few hours during low tide. Mussels, on the other hand, cannot tolerate tidal exposure even during low tide.

The best sites for culturing molluscs are therefore those that meet their biological requirements, including the following:

1. Seawater salinity range of 15-35 ppt.
2. Water depth of 1-10 m, and
3. Muddy bottom for mussels and hard rocky or coralline substrates for oysters.

In addition, the area for mollusc culture should be protected from strong water currents reaching three knots and should be accessible to source of seed, transport, and markets. Furthermore, the presence of local available stock in an area is a good indicator of its suitability for mollusc culture.

Countries which have successfully cultured bivalve molluscs have developed their own systems of culture which depend entirely on natural seed stock, which are either gathered from natural seed beds or collected using suitable materials for collecting seed from natural grounds.

In the Philippines, both natural and synthetic ropes have been used for spat collection. However, since natural ropes, which have been found to attract more larvae than synthetic polyethylene or polypropylene ropes, do not last long, natural fibrous materials like coconut coir are sometimes interwoven with synthetic nylon ropes to make them more attractive to the larvae.

The string seed collectors are submerged in the sea water for seed collection at the right time. They are hung on a collector rack, normally 12 strings along a distance of 1.8 m to hold about 1000 shells. Sometimes, strings are hung separately from each other at regular intervals; at others, three or four strings are put together for hanging to prevent branches from attaching to strings when they occur in large quantities.

Three principal methods of oyster culture are used in the Philippines and Japan:

1. Hanging method including rafts, longlines, simple hanging, and rocks;
2. Stake or stick method; and
3. Broadcast or sowing method.

In Japan, the earliest method used at the Hiroshima Prefecture, where oyster culture began in the 17th century, was the stick culture method. In 1927, the hanging method of culture was introduced which later developed into different variations, viz., the simple hanging method, raft method, and longline method, to suit different local conditions as culture grounds shifted from inner to outer parts of the bay to outer open seas.

The broadcast system is actually used throughout the world in places where the bottom of shallow bays is firm enough to support the materials used as collectors and for growing oysters. Oyster shells, stones, or other hard objects are scattered on the bottom in areas where setting or the attachment of oyster larvae is known to occur. The young oysters or spat are left in places attached to the collectors until they are large enough for harvest.

The stake method is usually applied in shallow areas with soft or muddy bottom, usually not more than 1 m deep during low tide. The stakes, usually bamboo trunks (whole or split), branches of mangrove trees, or concrete Y-shaped posts and other similar materials are staked on the sea bottom in rows spaced about 0.5 m apart, to serve as attachment for oyster spat.

The hanging method of oyster culture uses empty oyster shells or other material such as coconut shells as collectors. The collectors are strung on synthetic twine or heavy monofilament nylon, and placed about 10 cm apart by using bamboo tubes as spacers or by tying knots in the twine. The strings are hung from a platform or rack/tray made of bamboo or wooden splits or welded wire with wooden frame, and placed on wooden plots. Oysters detached from the collectors or those small oysters/seedlings which are separated from harvested stocks are cultured on the trays until they are big enough for the market.

Harvesting procedures vary with the culture method. Oysters grown on stakes or by hanging are removed from the stakes or ropes on shore or in a boat after the stakes/ropes are lifted out of the water. Those grown by broadcasting are usually collected at low tide.

Mussel farming makes extensive use of bamboos either as stakes or as floating rafts. The stake method, similar to that for oyster culture, is the most commonly used. The mussels are harvested by divers after 6-10 months when they reach a length of 5-8 cm.

Alternatively, mussels are grown on floating rafts which have the following advantages:

1. Faster growth;
2. Possibility of regular thinning and therefore higher production per unit area;
3. Possibility of transfer to other areas to prevent siltation; and
4. Ease of construction using more durable materials.

Mussels and oysters grown in waters contaminated by domestic and industrial wastes need to undergo depuration or cleansing, using artificially cleaned water or clean seawater from saltwater wells, to ensure satisfactory microbiological and chemical quality of the product. The depuration process flow and schematic diagram of a shellfish purification plant are shown in figure above.

Seaweed Farming

Seaweeds, aside from being used as food, are important sources of colloids or gels, such as agar, as well as minerals of medicinal importance such as iodine. Eucheuma, a red algae, is a valuable source of carrageenan, an important industrial compound used in stabilizing and improving the quality of a great number of products. Caulerpa lentillifera, a green algae, is economically important because it is a favourite and nutritious salad dish containing essential trace minerals such as calcium, potassium, magnesium, sodium, copper, iron and zinc. It is also known for its medicinal properties, being used as an anti-fungal agent and as a natural means for lowering blood pressure. Gracilaria, another red alga, is economically important in Taiwan (PC) for its agar extracts.

The culture of the seaweed Porphyra is believed to have started as early as between 1596 and 1614 in Hiroshima Bay utilizing pole and net devices originally installed to catch fish. At present, commercial seaweed culture is limited to five countries in East Asia, viz., Japan and Korea (which both grow mainly Porphyra, Undaria and Laminaria), China (Porphyra and Laminaria), Taiwan (PC) (Gracilaria and Porphyra), and the Philippines (Eucheuma spinosium, E. cottonii and Caulerpa lentillifera). Thirty-one species belonging to 18 genera and three divisions are presently cultured in these five countries, of which only three out of the 31 species are green algae.

In 1988, the estimated world seaweed production for use in the manufacture of carrageenan was nearly 68,000 t of dried seaweeds, of which nearly 66% was supplied by the Philippines and the rest by Indonesia, Chile and Canada. The bulk of the Philippine seaweed production consists of Eucheuma produced mainly in the southern part of the country in reef-protected coastal areas. Caulerpa is also successfully farmed in seawater ponds in Mactan, Cebu.

In Taiwan (PC), Gracilaria is cultured in ponds formerly used for milkfish, with Pingtung County alone accounting for 110 ha of the total 400 ha of Gracilaria ponds in Taiwan (PC) in 1974 and producing 1,000 t of dried Gracilaria seaweed.

In Japan, indoor facilities are used to obtain buds/seedlings for on-growing at sea. The facilities consist of 70-80 cm deep square or rectangular concrete tanks provided with illumination, a temperature control system, and ventilation.

The successful cultivation of seaweeds depends on four important factors:

1. Type of Seaweeds Used: The seaweeds cultured must be healthy and resistant to disease and breakage. They must be able to grow fast and give high yields during harvest. During processing, they must have high amounts of dry matter from which will be extracted high concentrations of carrageenan of high gel strength and viscosity.
2. Ecological Conditions of the Farm: The farm must be well-sited and fulfill the bio-ecological requirements of the culture species. In general, the presence of a particular seaweed species in an area is a good indicator of the suitability of that site for culture of the species under consideration.
3. Access to Sunlight: Seaweeds being cultivated need abundant sunlight for photosynthesis. Shading by other seaweeds and plants must be prevented by regular inspection and removal of the unwanted plants.
4. The Seaweed Farmer: The personality and dedication of the seaweed farmer is an important factor since the farmer must visit the farm regularly and carry out routine inspections. Some of the farmer's chores include shaking off silt and other foreign materials from the seaweeds, repairing broken lines, restoring uprooted stakes, and picking up drifting branches of seaweeds.

Trono and Ganzon-Fortes listed the following criteria for selecting good sites for Eucheuma in open waters and Caulerpa and Gracilaria in seawater ponds:

1. Unpolluted seawater supply.
2. Salinity of 30-35 ppt Eucheuma and Caulerpa and 8-25 ppt for Gracilaria.
3. Water temperature of 27-30 °C.
4. Moderate water movement of 20-50 m/min.
5. Water depth of 0.5-1 m at low tides and not more than 2-3 m at high tides, and
6. Firm bottom protected from strong waves for Eucheuma and muddy-loam bottom for Caulerpa ponds.

Seaweeds are grown using different types of planting material (vegetative cuttings, natural seeds, hatchery-reared seeds) and methods of culture (store planting, bottom culture, rope method, rope-concrete method, and pond culture either in monoculture or polyculture with milkfish, shrimp and crabs).

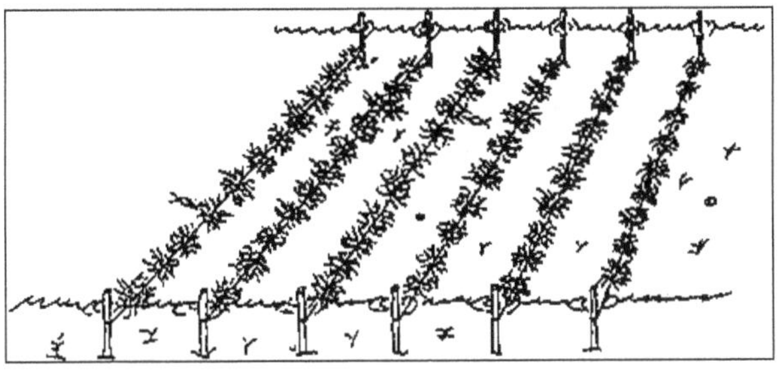

Three methods of Eucheuma culture practised in the Philippines .

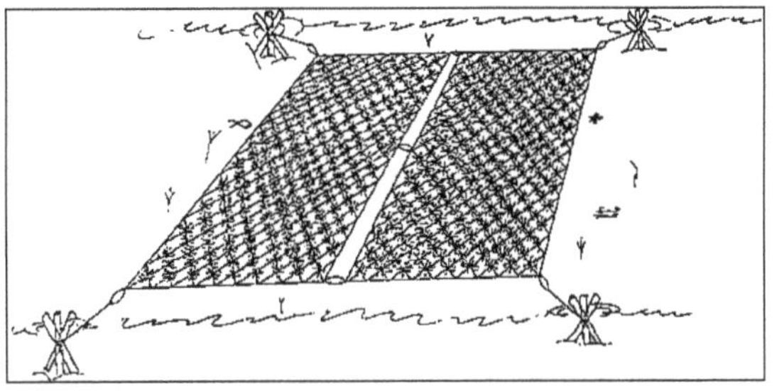

Three methods of Eucheuma culture practised in the Philippines.

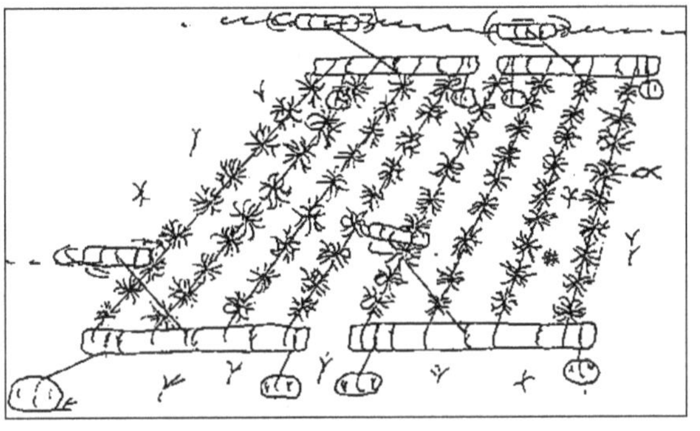

Three methods of Eucheuma culture practised in the Philippines.

In the Philippines, the monoline method of culture is the most popular and successfully used of these methods. The farming activities involved in monoline culture of Eucheuma species based on the Philippine experience are as follows:

1. Securing a license from the Bureau of Fisheries and Aquatic Resources (BFAR) prior to farming the area.
2. Preparing required materials needed for farm construction.
3. Clearing the area of sea grass, seaweeds, large stones and corals, and other foreign materials, followed by measuring it according to the proposed dimensions of the farm.
4. Wooden stakes are then driven into the bottom with the help of an iron bar and sledgehammer and arranged into 10 m rows at 1 m intervals. An 11 m nylon line is securely tied to one end of each stake about 0.5 m above the ground and then stretched to the corresponding opposite stake and tied securely. If the current is very strong, an additional row of stakes is placed in the middle to provide additional support.
5. Obtaining seedlings from the nearest source and transporting them to the farm site within the shortest possible time. During transport, the seedlings are protected from exposure to sun, wind, heat or rain. If the transport of seaweeds will take several hours, the seaweeds are kept damp during the trip and upon arrival at the farm, are immediately submerged in water.
6. Preparing the seedlings by tying bunches weighing about 50-100 g with soft 25 cm long plastic straw, and then tying these to monolines in the water at 20-25 cm intervals. The plants are allowed to grow to about 1 kg or larger before harvesting.

7. Building a farm house if drying of the harvested seaweeds is part of the operations. The farm house is built in or near the farm site so as not to waste time during post-harvest handling. The size of the farm house, which is designed to provide for drying and storage, will depend on the farmer's financial capacity and market commitments.
8. Maintaining planted seaweeds by inspecting them regularly while they are growing. Unwanted seaweeds which will compete with the Eucheuma for nutrients and sunlight are removed along with dirt and other foreign materials clinging to the seaweeds. Lost or broken Eucheuma are replaced.
9. Harvesting the whole plant and reserving select portions as seedlings for the next crop.
10. Sun-drying of the rest of the harvest by spreading these on a drying platform of bamboo slots initially lined with coarse fine-mesh nylon net. The seaweeds are freed of all foreign matter clinging to them.

During hot and sunny weather, it takes about 3-4 days to dry the seaweeds to a moisture content of about 30% or less. The dried materials are then packed in plastic sacks for storage in a dry place or for delivery to the buyer.

The pond culture of Caulerpa involves the following major steps:

1. Pond Construction: The pond is divided into manageable units measuring about 0.10-0.25 ha. The pond design allows for a flow-through system by providing each unit with its own supply and drainage gates. Water flows uniformly from the main gate to the secondary and exit gates during the draining and flooding process. Peripheral diversion dikes or canals along the landward edge of the pond are also built to divert run-off water from the ponds during the rainy season.

2. Planting: To facilitate planting activities, pond water is drained to a depth of about 0.3 m.

Caulerpa seedlings are obtained from the nearest source available and transported to the farm site within the shortest possible time.

The ponds are stocked at a rate of 1 000 kg seedlings/ha or 100 g/m2. A handful of seedlings is uniformly buried on one end at approximately 1 m intervals using a string as guide.

After planting, the pond water is gradually raised to a depth of 0.5-0.8 m or just until the plants can be seen from the surface of the water.

The newly planted seaweeds are inspected after a few days. Uprooted seaweeds are replaced and bare areas are replanted.

3. Pond Management: Water is changed daily or every other day to maintain adequate levels of nutrients. During the initial stages of growth, the seaweeds deplete the water of nutrients at a high rate and frequent water changes are needed to replenish lost nutrients and eliminate the need to fertilize. Water level is, however, carefully maintained to prevent the collapse of the dikes.

Unwanted seaweeds, sea grasses, and animals which will compete with the Caulerpa for nutrients are regularly weeded out.

The dikes and pond gates are inspected regularly to check for leakages, which are repaired immediately. This is vital, especially during the typhoon season.

The application of fertilizer may not be necessary as long as frequent water change is maintained. However, fertilization is resorted to when the stocks appear unhealthy and pale in colour, i.e., from light green to yellowish. When this happens, pond water is changed and fertilizer with a high nitrogen content is applied at the rate of 16 kg/ha by broadcasting or by suspending the fertilizer contained in several layers of plastic sack in strategic areas in the pond. The pond water is not changed in the next two to three days.

4. Harvesting: Two months after planting, the Caulerpa forms a uniform carpet on the pond bottom, a good indicator for harvest time.

About 75% of the crop is harvested by uprooting the Caulerpa from the mud and placing it on to a wooden raft.

About 25% of the original crop is left behind, uniformly spaced on the pond bottom to serve as seedstock for the next crop. This may be harvested after two to three weeks.

Harvested seaweeds are washed in clean sea water to remove mud and other dirt. The clean seaweeds are then placed in a basket or clean plastic sheets for further sorting and cleaning before packaging and immediate transport to the market.

Benefits and Importance of Aquaculture

Economic Benefits

1. Alternative food source: Fish and other seafood are good sources of protein. They also have more nutritional value like the addition of natural oils into the diet such as omega 3 fatty acids. Also since it offers white meat, it is better for the blood in reducing cholesterol levels as opposed to beef's red meat. Fish is also easier to keep compared to other meat producing animals as they are able to convert more feed into protein. Therefore, its overall conversion of pound of food to pound of protein makes it cheaper to rear fish as they use the food more efficiently.

2. Alternative fuel source: Algae are slowly being developed into alternative fuel sources by having them produce fuels that can replace the contemporary fossil fuels. Algae produce lipids that if harvested can be burn as an alternative fuel source whose only by products would be water when burnt.

Such a breakthrough could ease the dependency of the world on drilled fossil fuels as well as reduce the price of energy by having it grown instead of drilling petroleum. Moreover, algae fuel is cleaner and farmable source of energy, which means it can revolutionize the energy sector and create a more stable economy that avoids the boom-bust nature of oil and replaces it with a more abundant fuel source.

3. Increase Jobs in the market: Aquaculture increases the number of possible jobs in the market as it provides both new products for a market and create job opportunities because of the labor required

to maintain the pools and harvest the organisms grown. The increase in jobs is mostly realized in third world countries as aquaculture provides both a food source and an extra source of income to supplement those who live in these regions.

Aquaculture also saves fishermen time as they do not have to spend their days at sea fishing. It allows them free time to pursue other economic activities like engaging in alternative businesses. This increase in entrepreneurship provides more hiring possibilities and more jobs.

4. Reduce Sea Food Trade Deficit: The sea food trade in America is mainly based on trade from Asia and Europe, with most of it being imported. The resultant balance places a trade deficit on the nation. Aquaculture would provide a means for the reduction of this deficit at a lower opportunity cost as local production would mean that the sea food would be fresher. It would also be cheaper due to reduce transport costs.

Environmental Benefits

1. Creates Barrier against pollution with mollusc and sea weed: Molluscs are filter feeders while seaweed acts a lot like the grass of the sea. Both these organisms sift the water that flows through them as brought in by the current and clean the water. This provides a buffer region that protects the rest of the sea from pollution from the land, specifically from activities that disturb the sea bed and raise dust.

Also, the economic benefits of molluscs and sea weed can create more pressure from governments to protect their habitats as they serve an economic importance. The financial benefits realised provides incentive for the government to protect the seas in order to protect sea food revenue.

2. Reduces fishing pressure on wild stock: The practice of aquaculture allow for alternative sources of food instead of fishing the same species in their natural habitats. Population numbers of some wild stocks of some species are in danger of being depleted due to overfishing.

Aquaculture provides an alternative by allowing farmers to breed those same species in captivity and allow the wild populations to revitalize. The incentive of less labor for more gains pushes fishers to convert to fish farmers and make even more profit that before. It also allows them control of the supply of the fish in the market giving them the ability to create surplus stock or reduce their production to reap the best profits available.

Importance of Aquaculture

1. Sustainable use of sea resources: Aquaculture provides alternatives for fishing from the sea. Increase in demand for food sources and increase in globalization has led to increase in fishing. Yet, this has led fishermen to become selfish and overfish the desired or high-demand species. Through aquaculture, it provides both an alternative and opportunity for wild stocks to replenish overtime.

2. Conservation of Biodiversity: Aquacultures also protect biodiversity by reducing the fishing activities on wild stock in their ecosystems. By providing alternatives to fishing, there is reduced attack on the wild populations of the various species in the sea. Reduced action of fishing saves the diversity of the aquatic ecosystem from extinction due to overfishing.

3. Increased Efficiency, more resources for less effort: Fish convert feed into body protein more efficiently than cattle or chicken production. It is much more efficient meaning that the fish companies make more food for less feed. Such an efficiency means that less food and energy is used to produce food, meaning that the production process is cheaper as well. It saves resources and even allows for more food to be produced leading to secure reserves and less stress on the environment.

4. Reduced Environmental Disturbance: By increasing aquaculture, fish farming in specific, there is a reduced need for the fishing of the wild stock. As an outcome, it puts less stress on the ecosystem and equally reduces human interference. Actions of motor boats and other human influences such as the removal of viable breeding adult fish are all stresses put on the aquatic ecosystems and their discontinuation allows the ecosystem to flourish and find their natural balance.

References

- Aquaculture techtarget.com, Retrieved 2 March, 2019
- What-is-aquaculture-why-do-we-need-it: aquaculturealliance.org, Retrieved 10 May, 2019
- Aquaculture-characters-types-and-qualities, aquaculture: notesonzoology.com, Retrieved 4 January, 2019
- Aquaculture: fao.org, Retrieved 15 April, 2019
- Aquaculture-types-benefits-importance: conserve-energy-future.com, Retrieved 20 February, 2019

Chapter 2

Types of Aquaculture

The branch of aquaculture which deals with the cultivation of marine organisms in seawater is known as mariculture. Based on the organism cultivated, there are several types of aquaculture such as fish farming and shellfish aquaculture. The topics elaborated in this chapter will help in gaining a better perspective about these types of aquaculture.

Fish Farming

Fish farming is a form of aquaculture in which fish are raised in enclosures to be sold as food. It is the fastest growing area of animal food production. Today, about half the fish consumed globally are raised in these artificial environments. Commonly farmed species include salmon, tuna, cod, trout and halibut. These "aquafarms" can take the form of mesh cages submerged in natural bodies of water, or concrete enclosures on land.

According to the United Nations Food and Agriculture Organization, roughly 32% of world fish stocks are overexploited, depleted or recovering and need of being urgently rebuilt. Fish farming is hailed by some as a solution to the overfishing problem. However, these farms are far from benign and can severely damage ecosystems by introducing diseases, pollutants and invasive species. The damage caused by fish farms varies, depending on the type of fish, how it is raised and fed, the size of the production, and where the farm is located.

One significant issue is that—rather than easing the impact on wild populations—the farms often depend on wild fish species lower on the food chain, like anchovies, in order to feed the larger, carnivorous farmed species. It can take up to five pounds of smaller fish to produce one pound of a fish like salmon or sea bass. Overfishing of these smaller "forage" fish has repercussions throughout the ocean ecosystem.

As is the case with industrial animal farms on land, the fish are often housed in unnaturally crowded and cramped conditions with little room to move. Fish may suffer from lesions, fin damage and other debilitating injuries. The overcrowded and stressful conditions promote disease and parasite outbreaks—such as sea lice—that farmers treat with pesticides and antibiotics. The use of antibiotics can create drug-resistant strains of diseases that can harm wildlife populations and even humans that eat the farmed fish.

Escaped fish introduce yet another threat into the environment. Each year, hundreds of thousands of fish escape farms and threaten the genetic diversity and survival of native species. High stocking densities result in a significant amount of pollution from fish excrement and uneaten food, which in turn lead to poor water quality high in ammonia and low in oxygen. Outdoor fish farms can also attract predatory marine animals, such as sea birds and sea lions, who are sometimes poisoned or shot by aquafarmers for eating the fish.

Despite evidence to the contrary, it is still a common misconception that fish do not feel pain. Slaughter methods in the aquaculture industry are appalling. Little to no attention is given to the suffering of the animals and most are fully conscious during slaughter, which can take many minutes. Some species, such as salmon in the United States, are also starved for many days to empty the gut before they are sent to slaughter. Fish are most often not stunned and are killed by bleeding out, being hit on the head repeatedly, suffocating or freezing. In the US, as with many other countries, there are no regulations to ensure the humane treatment of fish.

Years of unregulated and underreported catches of bluefin tuna in the Mediterranean Sea and Atlantic Ocean are threatening the existence of this severely overfished species. To meet the high and growing demand for sushi in Japan and elsewhere, ranching of bluefin tuna is becoming a popular industry and is exacerbating the problem. Fisherman use longlines and purse seines to catch the tuna before they reach breeding age and have time to reproduce. They are then kept in seafarms for 3–6 months and fattened with thousands of pounds of smaller wild-caught fish before being killed and exported.

Advantages of Fish Farming Business

There are many advantages of starting commercial fish farming business. Here we are describing the main advantages of fish farming business.

- According to the demand, commercial fish farming business allows for large supplies of fish. Catching fish from the wild can't always fulfill the consumer's demand. In such cases commercial fish farming can meet up this demand.
- Fish can be raised in tanks until they are ready for marketing, and they don't require the extensive capture of wild fish. Thus commercial fish farming helps to preserve natural ecosystems.
- Compared to the wild brethren, some farm raised fish species are more nutritious. Fish are usually fed a wide variety of protein and nutrient enriched foods or pellets on commercial fish farms. So farm fish become more healthier than the wild fish.
- Various types of fish species are available throughout the world. So you can choose your desired species for your fish farming business.
- Fish are very popular as food around the world. So there is an established fish market, and you don't have to worry about marketing your products.
- You can start fish farming business in both large or small scale production purpose.
- If you don't have enough capital for starting this business, then you can apply for bank loans. Many banks will allow loans for starting this business commercially.
- Fish farming business is a great source of employment. More than 1 billion people around the world depend on fish as their primary protein source. And most of this people are directly or indirectly involved with fish products or fish farming business. As a result, fish farming creates a great income and employment source for the people. Global fish exportation business is now earning more money every year, than any other food commodity.

- Even you can meet up your daily family nutritional demands through small scale fish farming in tanks or ponds.

Fish Farming Business

For maintaining a profitable fish farming business, you have to go through some step by step process. The steps for starting a fish farming business includes selecting suitable farm land or area, fish farm type (cage, tank or pond), cage or pond construction, selecting fish species, feeding, care & management, harvesting and marketing. We are shortly describing all the steps below. For running a successful fish farming business, follow every steps very carefully.

Select a Suitable Farm Land/Area

Selecting a suitable land or area is very important for starting commercial fish farming business. All the areas are not suitable for profitable business. And some areas have plenty of natural resources, which are very effective for fish farming business. Especially coastal areas and the areas near big rivers or stream are very suitable for establishing fish farming business. Consider the followings while selecting land or areas for your business.

- Select relatively level land and avoid steeply sloped lands.
- Consider your future business plan, while selecting the land. It will be better if you can select a large piece of land, where you can perform all types of necessary farm activities.
- Avoid flooding and polluted areas, because flooding area can harm your business seriously.
- Don't select fish farming land near the crop fields. Farmers usually use a lots of fertilizers and pesticides in their field for better production. These chemicals get mixed with water and the polluted water can affect your fish farm.
- It will be better, if your selected land become slightly lower than the main water source. It also help to reduce the cost of filling your farm land with water. Natural gravity will fill the land without any cost.
- Ensure good transportation system is available in your selected area. Good transportation system will be very effective for marketing your products and purchasing necessary commodities from the market.

Type of Fish Farm

There are numerous specific type of fish farms in both intensive and extensive fish farming system. You can start fish farming business by choosing any type. You can choose cage system, tank system or pond system. In cage system, make a suitable cage and place it in lakes, ponds, bayous or oceans and start feeding the fish until they reach marketing age. In case of raising fish in tanks, make a or a few tanks and stock fish there. Here we are describing more about fish farming in pond system.

Pond Design and Construction

Construct a suitable pond, after selecting your farm area. Before constructing, make a good design and make the pond according to your desired design. While designing the pond, ensure the availability of all types of necessary facilities for maintaining a profitable fish farming business. Although the design of a pond depends on the fish species you intend to raise and your location. You can consult with your nearest fisheries institute to learn more about specific pond design for specific fish species. Always try to maintain a good environment in the pond. Good environment helps to live and grow the fish well, and it directly involved with better production and maximum profits.

Suitable Fish Species

Selecting suitable fish species is very important for maximizing profits form fish farming business. Select those breeds, which have a huge demand and high price in your local market. The most important fish species used in fish farming throughout the world are carp, salmon, tilapia and catfish. All of these fish species have many varieties and suitable for farming in all types of agro-climatic conditions. Select the fish species for farming, depending on your local facilities, demand and price.

Feeding is the most important part of commercial fish farming business. Always try to provide your fish high quality and nutritious food. High quality food not only ensures maximum production but also help to keep the fish healthy. So provide your fish supplementary fish feed along with natural food. There are various types of prepared commercial fish feed available in the market for specific fish species. You can purchase those commercial feed from the market or prepare it by your own. Learn more about preparing supplementary fish feed, if you want to prepare it by your own. Don't forget to add all types of necessary nutrient elements, such as vitamins, minerals, salt etc. Feed your fish several times a day, depending on the fish species.

Care and Management

Always try to provide your fish fresh and nutritious foods. It will be better, if you can change water from the pond occasionally. If not possible, then you can use some chemicals according to the suggestion of an expert. Monitor the health of your fish on a regular basis. Do all your necessary farm tasks timely. Keep the pond environment clean and suitable for proper growth. Test the water and soil quality of your pond on a regular basis. Always stock some necessary drugs on your farm. Prevent all types of predators, including frogs, snakes etc.

Harvesting

After a certain period, the fish become suitable for harvesting. Although this time depends on the fish species. Start harvesting, when a major numbers of fish reach marketing age. You can use net for harvesting fish or by removing water from the pond. Try to harvest during morning or afternoon, when temperature is low. After harvesting, send the fish to the market as soon as possible.

Marketing

Marketing is the easiest step of fish farming business. There are numerous markets available where you can sell your products. And all types of fish have a huge demand in the market. After harvesting, you can easily sell the fish at any of your nearest local market. Even there are many companies available who export fish to the foreign countries. So don't worry about marketing the products, just focus on the other steps.

In a word, commercial fish farming business is really very profitable and a good source of earning livings. If you intend to join this venture, then visit some fish farms in your area and try to have some practical knowledge.

Impacts and Issues

Fish farming has become a very contentious practice, for a number of environmental reasons and for the adverse health effects it has on the farmed fish and possibly other species, including humans.

In a fish farm, the concentration of fish far exceeds that found in schools of fish in the wild—50,000 or more fish in an area of several acres in volume—with the possible exception of the spawning runs of west coast salmon. These crowded conditions reduce the free-swimming volume of each fish to about that of the average household bathtub. In such crowded conditions, the fish bump and rub against each other in the boundaries of the pens, which can produce cuts and scrapes. This increases the likelihood of infection and the development of diseases.

Species of sea lice that parasitize Coho and Atlantic salmon are especially troublesome. The sea lice attach to the fish and feed on tissue, which creates lesions and causes fluid loss from the affected fish. The confined fish become ill and can die. In addition, the sea lice can spread to wild salmon in the seas around fish farms when farmed salmon escape from the confinement, and also when the lice are washed away from the fish farm into the surrounding water. A 2001 survey of wild juvenile salmon migrating past fish farms in British Columbia found many more sea lice on the juveniles that had passed the farms than on those who had not yet passed by the facilities.

The escape of fish from fish farms is not a trivial and isolated event. Rips and breaks in the pen material and buffeting of the pens by storm-driven waves can lead to the escape of fish. In some cases, pens are designed with a net lid to reduce this possibility. Sometimes only a few fish escape. But mass escapes have occurred. For example, in January 2002, over 8,000 fish escaped from a fish farm in Clayoquot Sound, British Columbia. Worldwide in 2004, an estimated 2 million farmed fish escaped to the wild.

Once in the wild, the farmed fish have the potential to transfer disease to the wild population. An article in a December 2007 issue of Science documented declines in the population of wild Pacific salmon related to their decimation from sea lice transferred from farm populations of Atlantic salmon. The situation is so dire that the natural population could be reduced by 99% by 2015, which would be an economic disaster for the traditional salmon fishery and those employed by the fishery.

Antibiotics can be supplied to the food in an effort to control infections. As with land-bound factory farms, this practice encourages the development of antibiotic resistance among the surviving bacteria. These hardier varieties of the bacteria may pose a health hazard not only to the farmed fish, but to wild fish populations and to humans.

A fish farm releases a great amount of untreated sewage to the surrounding water. A study done in Clayoquot Sound calculated that the 700,000 fish housed in the facility that is the size of three football fields generate the daily equivalent amount of sewage produced by 150,000 people.

Elsewhere, the situation is not better. In China, for example, which produces approximately 70% of the world's farmed fish, fish farms can be concentrated together around large ponds. Analysis of the pond waters has revealed the presence of pesticides and other agricultural run-off, antibiotics, and cancer-causing compounds. The result is both an environment and food safety problem. In late 2007, the United States and China signed an agreement to permit more monitoring of the farms and the safety of the exported products.

Sustainability of Fish Farming

Fish farming in Marseille, France.

We should eat fish at least twice a week, since fish are high in protein, low in saturated fats and rich in omega-3 fatty acids. Global per capita fish consumption has almost doubled from the 1960s to 2012. And today, about half of all the seafood destined for human consumption is produced through fish farming, also called aquaculture.

The Food and Agriculture Organization of the U.N. (FAO) projects that by 2030, fish farming, one of the fastest growing methods of producing food in the world, will be responsible for almost two-thirds of the fish we eat.

The most common type of aquaculture is farming in net pens or cages anchored to the sea floor in the ocean near the coast. There are also closed systems of tanks or ponds that float on water or operate on land. The FAO estimates that over 600 aquatic species are produced globally in a variety of aquaculture systems using freshwater, brackish water or salt water.

In the U.S., over 91 percent of the seafood we eat is imported. China is the largest exporter of fish globally, the third largest importer of fish and the biggest aquaculture producer. As early as 2500 B.C., the Chinese practiced fish farming, putting carp in rice paddies where they ate insects and weeds, fertilized the rice, and then were eaten. Today, 88 percent of the world's aquaculture comes from Asia.

While fish are known for their omega-3 fatty acid benefits, they do not actually produce omega-3s themselves. Microscopic algae that live in fresh or salt water or sediments produce them. Herbivorous fish and forage fish like sardines, anchovies and herring obtain omega-3s by eating the microalgae. Larger carnivorous fish such as salmon or sea bass then eat the forage fish. Because salmon and other popular carnivorous fish need omega-3s to grow, 30 to 50 percent of the fish feed traditionally used for these species consists of fishmeal (ground fish) and fish oil. Over 50 percent of the world's fish oil is used in feed for farmed salmon.

Anchovies

This is one reason that fish farming has a reputation as being unsustainable. In 1997, it took almost 3 tons of forage fish to produce one ton of salmon. A third of the global fish harvest still goes toward making fish meal and fish oil. As a result, forage fish are being overfished, and some populations have crashed, which has implications for the entire food web since larger fish depend on them for food.

Most fish farming methods are harmful to the ecosystem in other ways as well.

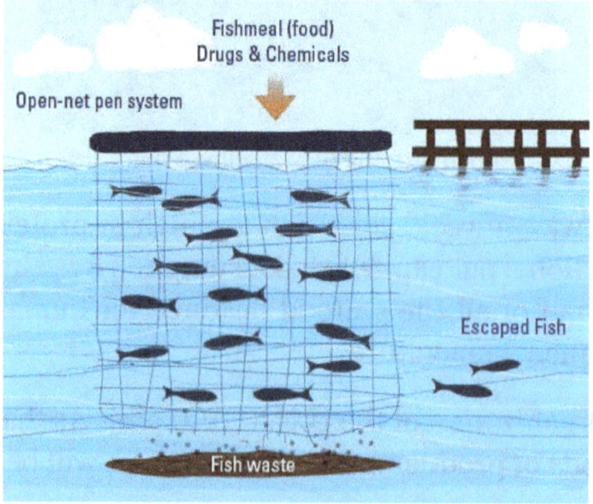

Fish waste and left over food spill out from nets into the ocean, causing nutrient pollution. This may lead to oxygen depletion in the water, which can stress or kill aquatic creatures. In addition, antibiotics or pesticides used on farmed fish can affect other marine life or human health. These nutrients and chemicals also sink to the ocean floor, where they may impact the biodiversity there.

Fish crowded together in nets or pens are more susceptible to stress, which can foster disease and parasites that may then spread to wild species.

Farmed fish sometimes escape into the ocean, breeding with wild species and affecting the population's overall genetic diversity.

Fish farming in tanks.

Land-based closed systems minimize the amount of waste and nutrients expelled to the environment, eliminate fish escapes and limit the spread of disease; but pumping the water through them requires a great deal of energy, and the wastewater must still be disposed of properly.

Fish farming has also resulted in land conversion for feed and the destruction of ecosystems. To grow the soybeans used in herbivorous and other fish feed, vast areas of South America, almost 4 million hectares of forest, are razed yearly and converted to agricultural land. Growing the soybeans also depends on water availability.

Shrimp farming, which is usually done in salty coastal waters, has been responsible for the destruction of 38 percent of the world's mangroves. Mangroves have critical ecological functions including providing food and habitat for many species, preventing erosion, sequestering carbon, and offering protection from storms. Moreover, many shrimp ponds accumulate shrimp waste, antibiotics and pesticides, and without mangroves to filter them, eventually become unusable.

"Most aquaculture operations are far more sustainable than animal food production," "Aquaculture gets a bad rap because we see it as an alternative to natural environments like coasts that haven't been developed, as opposed to comparing it with agriculture. Actually aquaculture is a much more sustainable way to get protein than agriculture."

"A lot of damage is being caused by aquaculture, but it's getting much better," "People are getting educated, and it's becoming unacceptable to farm fish in a way that's harmful to the environment."

How Fish Farming is Becoming more Sustainable

One strategy involves moving aquaculture out into the open ocean where the water is pristine and currents are strong and steady enough to continually flush the farms of fish waste and pests such as sea lice. The open ocean also provides farmed fish with more consistent salinity and temperature. This means they are less stressed and less vulnerable to disease, which promotes better growth and minimizes the need for antibiotics or vaccines.

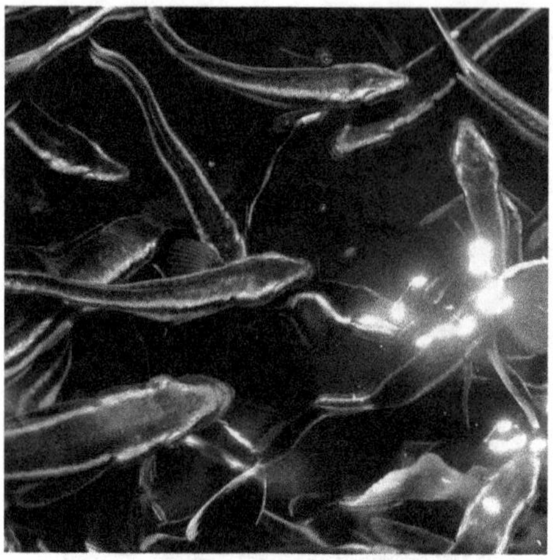

Cobia

Open Blue Sea Farms grows cobia (related to remoras) in the largest open ocean farm in the world. After the fish develop in the hatchery, they spend 14 months in huge submersible pens deep below the surface in clean ocean waters seven miles off the coast of Panama.

Steve Page of Ocean Farm Technologies co-designed the Aquapod, a geodesic dome large enough to accommodate several hundred thousand fish. Kampachi Farms off the coast of Hawaii is using the Aquapod to grow kampachi (related to yellowtail), after successfully mounting the Velella Project. The project researched the viability of farming fish in an Aquapod tethered to a drifting boat in deep water. It had no measurable impact on the environment. Because satellite communication was not robust enough to handle the remote controls needed to manage the drifting Aquapod, however, Kampachi Farm is now using an Aquapod attached to a barge out in the ocean. In the future, Aquapods could potentially be equipped with propellers and a GPS system, and used to transport juvenile fish to arrive at their destination with the fish ready to harvest.

On land, some fish farms are using recirculation systems to recycle their water. Recirculation systems use 100 times less water per kilo of fish than traditional land-based systems. In addition, the water quality can be monitored continuously, which lessens the risk of disease and the need for antibiotics.

Denmark is a leader in recirculation system aquaculture. Hallenbaek Dambrug raises rainbow trout while recirculating over 96 percent of its water. The discharge wastewater is filtered, and the sludge used for biogas or fertilizer. The discarded water is treated to remove nitrate.

Anadramous fish like salmon and trout are born in fresh water then migrate to the ocean, returning to freshwater to spawn. Salmon and trout are typically raised in fresh water until they are mature enough to migrate to salt water, where they are farmed in sea cages. But some new recirculation systems allow these fish to spend their entire life on land by alternating fresh and salt water environments through controlling the water chemistry.

Scientists at the University of Maryland Department of Marine Biotechnology developed a recirculation system that facilitates predictable reproduction in farmed fish, one of the main challenges of aquaculture. The system recycles 99 percent of its water, filters waste through microbial communities and produces methane as a biofuel. By changing water temperature, lighting and salinity levels, and then feeding the fish a pellet that mimics a hormone prompting reproduction, the scientists have been able to get the fish to reproduce in predictable cycles.

RDM Aquaculture, an indoor saltwater shrimp farm in Indiana, has recycled the same water for five years, produces zero waste and uses no chemicals. Its "heterotrophic biofloc system" allows all organic matter—shrimp waste, bacteria, microalgae, shrimp shells and dead shrimp—to remain in the water. The shrimp eat what they need to eat and the bacteria feed on their waste.

An Integrated multi-trophic aquaculture site at Cooke Aquaculture Inc. in the Bay of Fundy, Canada: salmon cages (left), mussel raft (right foreground) and seaweed raft (right background).

A Dutch and Vietnamese industry group along with universities and research organizations is designing a "nutritious-system" concept of aquaculture in Vietnam that utilizes the natural ecosystem of a pond to farm fish or shrimp and get rid of the waste. The project studies how omega-3 fatty acids are produced and will determine the right balance of algae and bacteria to ensure the best water quality, nutrition for fish and shrimp, and decomposition of waste.

The rising prices of fish feed and the environmental impacts of over-exploiting forage fish for feed and fish oil have led to an increase in the farming of herbivorous fish (such as carp and tilapia) and omnivorous fish (barramundi) that require much less fishmeal to produce protein. Meanwhile, research is also ongoing to find alternatives to fishmeal feed or ways to make it more sustainable.

New Kinds of Feed

- Kampachi Farm has experimented with fish diets supplemented with soybeans and plant waste, and replacing fish oil with microalgae and yeast products. In 2013, Kampachi tested three feeds containing no fishmeal at all, and found them all comparable to the standard diet. Researchers at the University of Maryland Center for Environmental Science developed fish feed made completely with corn, wheat and soy. The fish oil was replaced by fatty acids from algae, amino acids and soybean or canola oil. PCB and mercury levels in the fish were 100 times less than those in fish eating fishmeal feed. This is because fishmeal and fish oil in feed can transfer environmental pollutants to farmed fish, while feed made of ingredients from vegetation can reduce them.

- British scientists genetically modified camelina sativa, a plant known for its seed oil, with synthesized genes from algae, enabling the plant to produce omega-3s that successfully replaced fish oil in fish feed; the salmon thrived.
- A Texas A&M University scientist is using distillers' dried grains with solubles, nutrient-rich grains made in ethanol production, as a cheap source of protein in shrimp feed. He has successfully substituted sorghum and corn distillers' dried grains for 10 percent of the protein in shrimp feed.
- Calysta, a California based company, is developing protein for feed using bacteria that are fermented and fed methane gas, in a process similar to making beer or bread. The product, called FeedKind, is a natural high-quality protein fishmeal replacement.
- Researchers from Wageningen University in the Netherlands are experimenting with insects as a new source of omega-3 fatty acids. The scientists extracted the naturally produced oil from a variety of insects and are researching the breeding, optimal diet and processing of the insects for oil. A 2014 FAO paper concluded that insect meal could replace between 25 and 100 percent of soymeal or fishmeal in fish diets with no adverse effects.

Aquaculture in Fuzhou, China.

In 2010, only 36 percent of fishmeal came from the trimmings and waste (heads and innards) of fish fillets, which are usually discarded. China increasingly relies on wild-caught fish for fishmeal and fish oil. A Stanford University study found that using the waste from seafood processing plants, and adding algae or ethanol yeast to boost the protein content (since waste has less protein than wild-caught fish), could replace half to two-thirds of the current fishmeal used in Chinese aquaculture.

Finding the best formulas for fish feed also means trying to achieve the lowest feed conversion ratio—the amount of feed given in relation to the amount of weight gained by the fish. For example, tilapia can produce a pound of protein on less than a pound of food, while salmon require a pound and a half of food to produce a pound of protein.

A tilapia fish farm in Honduras.

"It depends on how you farm them—the right stock density, water quality, and nutrient-rich food translate into converting well to protein." "And the lower the food conversion ratio, the more profitable the fish farm, because food is so expensive."

As the global population grows to 9 billion by 2050, more people enter the middle class, and fisheries are overexploited, fish farming will be critical to supplying the protein the world needs.

Making Sure of the Sustainability of Aquaculture

Because there are a number of different international and national certification schemes for aquaculture, the FAO developed technical guidelines for aquaculture certification and an evaluation framework. But while environmental impact assessments and certification are required for many large fish farms, they are not required for small farms, many of which are unsustainable. Regulations governing responsible aquaculture development in many countries are weak.

- The Global Aquaculture Alliance developed the voluntary Best Aquaculture Practices Certification. The standards address environmental and social responsibility, animal welfare, food safety and traceability.
- The Aquaculture Stewardship Council, founded by the World Wildlife Fund and the Dutch Sustainable Trade Initiative, also aims to make fish farming environmentally sustainable and socially responsible. Its goal is to be the global leader in certifying responsibly farmed seafood and managing global standards for sustainable aquaculture.
- The World Wildlife Fund, the FAO, the World Bank, and others formed the Shrimp Aquaculture and the Environment Consortium to adopt international principles for responsible shrimp farming.
- The SNV Netherlands Development Organization and the International Union for Conservation of Nature launched the Mangroves and Markets project in Cà Mau, Vietnam, to promote sustainable shrimp farming. The project provides training to farmers on breeding and marketing sustainably certified shrimp, promotes the replanting of mangrove forests, and helps shrimp farmers get certified in carbon markets. These shrimp farms must have 50 percent mangrove growth or more.

Mariculture

Mariculture is the cultivation and harvest of marine flora and fauna in a controlled saltwater environment. Sometimes called marine fish farming, marine aquaculture, or aquatic farming, mariculture involves some degree of human intervention to enhance the quality and/or quantity of a marine harvest. This may be achieved by feeding practices, protection from predators, breeding programs, or other means.

Fish, crustaceans, salt-water plants, and shellfish may be farm raised for bait, fishmeal and fish oil production, scientific research, biotechnology development, and repopulating threatened or endangered species. Ornamental fish are also sometimes raised by fish farms for commercial sales. The most widespread use of aquaculture, however, is the production of marine life for human food consumption. With seafood consumption steadily rising and overfishing of the seas a growing global problem, mariculture has been hailed as a low-cost, high-yield source of animal-derived protein.

According to the Fisheries Department of the United Nations' Food and Agriculture Organization (FAO), over 33 million metric tons of fish and shellfish encompassing 220 different species are cultured (or farmed) worldwide, representing an estimated $49 billion in 1999. Pound for pound, China leads the world in aquaculture production with 32.5% of world output. In comparison, the United States is only responsible for 4.4% of global aquaculture output by weight. Just 7% of the total U.S. aquatic production (both farmed and captured resources) is attributable to aquaculture (compared to 62% of China's total aquatic output).

In the United States, Atlantic salmon and channel catfish represent the largest segments of aquaculture production (34% and 40%, respectively, in 1997). Though most farmed seafood is consumed domestically, the United States imports over half of its total edible seafood annually, representing a $7 billion annual trade deficit in 2001. The Department of Commerce launched an aquaculture expansion program in 1999 with the goal of increasing domestic seafood supply derived from aquaculture production to $5 billion annually by the year 2025. According to the U.S. Joint Subcommittee on Aquaculture, U.S. aquaculture interests harvested 842 million pounds of product at an estimated value of $987 million in 1999.

To encourage further growth of the U.S. aquaculture industry, the National Aquaculture Act was passed in 1980 (with amendments in 1985 and 1996). The Act established funding and mandated the development of a national aquaculture plan that would encourage "aquaculture activities and programs in both the public and private sectors of the economy; that will result in increased aquacultural production, the coordination of domestic aquaculture efforts, the conservation and enhancement of aquatic resources, the creation of new industries and job opportunities, and other national benefits."

In the United States, aquaculture is regulated by the U.S. Department of Agriculture (USDA) and the Department of Commerce through the National Marine Fisheries Service (NMFS), National Oceanic and Atmospheric Administration (NOAA). State and local authorities may also have some input into the location and practices of mariculture facilities if they are located within an area governed by a Coastal Zone Management Plan (CZMP). Coastal Zone Management Plans, as authorized by the Coastal Zone Management Act (CZMA) of 1972, allow individual states to determine the appropriate use and development of their respective coastal zones. Subsequent amendments to the CZMA have also made provision for states to be eligible for federal funding for the creation of state plans, procedures, and regulations for mariculture activities in the coastal zone.

Tilapia, catfish, striped bass, yellow perch, walleye, salmon, and trout are just a few of the fresh and salt-water finned fish species farmed in the United States. Crawfish, shrimp, and shellfish are also cultured in the U.S. Some shellfish, such as oysters, mussels, and clams, are "planted" as juveniles and farmed to maturity, when they are harvested and sold. Shellfish farmers buy the juvenile shellfish (known as "spat") from shellfish hatcheries and nurseries. Oysters and mussels are attached to lines or nets and put in a controlled ocean environment, while clams are buried in the beach or in sandy substrate below low tide. All three of these shellfish species feed on plankton from salt water.

But just as overfarming takes a toll on terrestrial natural resources, aquaculture without appropriate environmental management can damage native ecosystems. Farmed fish are raised in open-flow pens and nets. Because large populations of farmed fish are often raised in confined areas, disease spreads easily and rapidly among them. And farmed fish often transmit sea lice and other parasites and diseases to wild fish, wiping out or crippling native stock. Organic pollution from effluent, the waste products from farmed fish, can build up and suffocate marine life on the sea floor below farming pens. This waste includes fish feces, which contributes to nutrient loading, and chemicals and drugs used to keep the fish disease free and promote growth. It also contains excess fish food, which often contains dyes to make farmed fish flesh more aesthetically analogous to its wild counterparts.

Farmed fish that escape from their pens interbreed with wild fish and weaken the genetic line of the native stock. If escaped fish are diseased, they can trigger outbreaks among indigenous marine life. Infectious Salmon Anemia, a viral disease that has plagued fish farms in New Brunswick and Scotland in the 1990s, was eventually found in native salmon. In 2001, the disease first appeared at U.S. Atlantic salmon farms off the coast of Maine.

The use of drugs in farmed fish and shellfish intended for human consumption is regulated by the U.S. Food and Drug Administration (FDA). In recent years, antibiotic resistance has been a growing issue in aquaculture, as fish have been treated with a wide array of human and veterinary antibiotic drugs to prevent disease.

The commercial development of genetically-engineered, or transgenic, farmed fish is also regulated by FDA. As of May 2002, no transgenic fish had been cleared by FDA for human consumption. The impact transgenic fish may have on the survival and reproduction of native species will have to be closely followed if and when commercial farming begins.

As mandated by the 1938 Mitchell Act, the NMFS funds 25 salmon hatcheries in the Columbia River Basin of the Pacific Northwest, the largest federal marine fishery program in the United States. These aquaculture facilities were originally introduced to assist in repopulation of salmon stocks that had been lost to or severely hampered by hydroelectric dam projects. However, some environmentalists charge that the salmon hatcheries may actually be endangering wild salmon further, by competing for local habitat and weakening the genetic line of native species.

Without careful resource management, aquaculture may eventually take an irreversible toll on other non-farmed marine species. Small pelagic fish, such as herring, anchovy, and chub, are captured and processed into fish food compounds for high-density carnivorous fish farms. According to the FAO, at its current rate, fish farming is consuming twice as many wild fish in feeding their domestic counterparts as aquaculture is producing in fish harvests—an average of 1.9 kg of wild fish required for every kilogram of fish farmed.

No matter how economically sound mariculture has been, it has also led to serious habitat modification and destruction, especially in mangrove forests. In the past twenty years, the area of mangrove forests have dwindled by 35% worldwide. Though some of that loss is due to active herbicide control of mangroves, their conversion to salt flats, and the industrial harvesting of forest products (wood chips and lumber), mariculture is responsible for 52% of the world's mangrove losses.

Mangrove forests are important to the environment because these ecosystems are buffer zones between saltwater areas and freshwater/land areas. Mangroves act as filters for agricultural nutrients and pollutants, trapping these contaminants before they reach the deeper waters of the ocean. They also prevent coastal erosion, provide spawning and nursery areas for fish and shellfish, host a variety of migratory wildlife (birds, fish, and mammals), and support habitats for a number of endangered species.

Shrimp farming, in particular, has played a major role in mangrove forest reduction. Increasing from 3% to 30% in less than 15 years, commercial shrimp farming has impacted coastal mangroves profoundly by cutting down mangrove trees to create shrimp and prawn ponds. In the Philippines, 50% of the mangrove environments were converted to ponds and between 50% and 80% of those in Southeast Asia were lost to pond culture as well.

Shrimp mariculture places high demands on resources. It requires large supplies of juvenile shrimp, which can seriously deplete natural shrimp stocks, and large quantities of shrimp meal to feed them. There also is considerable waste derived from shrimp production. This can pump organic matter and nutrients into the ponds, causing eutrophication, which causes algae bloom and oxygen depletion in the ponds themselves or even downstream. Many shrimp farmers need to pump pesticides, fungicides, parasiticides, and algicides into the ponds between harvests to sterilize them and mitigate the effects of nutrient loading. Shrimp ponds also have extremely short life spans, usually about 5–10 years, forcing their abandonment and the cutting of more mangrove forests to create new ponds.

Mariculture also limits other marine activities along coastal waters. Some aquaculture facilities can occupy large expanses of ocean along beaches which become commercial and recreational no-fish zones. These nursery areas are also sensitive to disturbances by recreational activities like boating or swimming and the introduction of pathogens by domestic or farm animals.

Shellfish Aquaculture

Unlike all other forms of marine aquaculture, commercially grown bivalve shellfish have been identified as the only sustainable form of aquaculture that has no negative impact on the environment.

The delicious oyster may be key to sustainable aquaculture and clean coastlines.

As the demand for seafood continues to surpass supplies of wild-caught fish and shellfish, marine aquaculture is becoming recognized as the only serious solution for feeding a future global population of 9 billion. However, critics contend conventional aquaculture has numerous challenges from caging fish in farms that generate waste from feces and unconsumed commercial feed. These wastes can carry disease and the phosphate and nitrates in the mix can cause Harmful Algal Blooms (HABs) that suck oxygen from the water, leaving it uninhabitable. Conventional near-shore cages have become excessively dependent upon pesticides and antibiotics to combat diseases that are rampant in highly concentrated farming conditions—not unlike industrial-scale hog, poultry, and cattle farming on land.

Shellfish aquaculture operations actually improve water quality by filtering out pollutants, sediments, and nutrients from the water column. Furthermore, "open ocean" farming is a new paradigm challenging traditional shellfish cultivation in bays and estuaries. The swift currents and upwelling supply ample food to promote faster growth rates and suspended open ocean long-lines prevent predation and parasites that impact shellfish in calmer shore waters. Low salinity and heavy siltation following torrential rains cause mortalities for shellfish harvested from congested and contaminated bays and estuaries leading to quality problems.

Shellfish Ecosystem Services

Filter-feeding bivalve shellfish oysters, mussels, clams and scallops are successfully farmed across the globe as a sustainable food source while also enhancing the marine environment. Incredibly, an adult shellfish can filter up to 50 gallons of water a day removing suspended solids from surrounding waters, thereby increasing water clarity allowing sunlight to penetrate for enabling sea grass growth and the foundations of the marine food chain to flourish. Beds of bivalve shellfish provide ecosystem services by naturally filtering silt and also removing bacteria, viruses and nutrients from the water.

The Nature Conservancy released a report Shellfish Reefs at Risk stating: "Centuries of intensive fisheries extraction exacerbated by more recent coastal degradation have put oyster reefs near or past the point of functional extinction worldwide. Globally, 85 percent of reefs have been lost, making oyster reefs the most severely impacted marine habitat on the planet".

Bivalve shellfish are no longer the global ecosystem engineers and enablers of prosperous habitats for other species. Enormous oyster beds in the United States' largest estuary, the Chesapeake Bay, could once filter its entire volume of water (about 19 trillion gallons) in a week. Today, it would take the nearly decimated remaining bay oysters more than a year to perform such services.

Harmful Algae Blooms

Several decades ago relatively few countries appeared to be affected by HABs but now most coastal countries are threatened. The causes of this global expansion are debated with possible explanations ranging from natural mechanisms of species dispersal to a host of human-related phenomena such as pollution in the form of nitrogen and phosphorus. These blooms form massive blankets of slime on the water's surface that precipitate bacteria growth, deplete oxygen, and kill much of the life in the water below.

HABs have been spreading and increasing along the coast of China in the past two decades, causing damage to the marine environment and posing a threat to human health. Chinese government and concerned scientists started to pay greater attention to HABs after the especially devastating bloom in the Bohai Sea, in August 1977. This event covered an area of 560 square kilometers and lasted 20 days, causing a mass mortality of fish and resulting in great losses to the local fishery. With increased awareness of the issue and economic development along the coast of China, HABs numbers have increased dramatically each year. It is clear that HABs have been increasing rapidly along the Chinese coast since the 1970s, and that they occur more frequently along the south coast than along the north coast. Thus, at least 322 documented HAB events have occurred from 1952-1998 in Mainland China.

A specific program for HAB monitoring along the Chinese coast, sponsored by the State Oceanic Administration, was initiated in 2001. At the present time, however, Mainland China, Hong Kong and Taiwan each utilize different systems for monitoring HABs. Since China has the largest mariculture operation in the world, future research efforts on the relationship between shellfish mariculture activity and HAB occurrences in key coastal areas, would foster and maintain the sustainable development of both shellfish mariculture operations and healthy marine ecosystems along the Chinese coast.

Scientists calculate that bacteria in sediment around bivalve beds biologically remove at least 20 percent of the nitrogen in wastes through the same process used in modern wastewater treatment plants. While most of the nutrients filtered from the water by shellfish are recycled back into the water column, the flux of undigested plant matter into the sediments stimulates bacterial processes known as denitrification. This process of turning fertilizer ammonia into nitrate and then into harmless nitrogen gas allows its escape into the atmosphere instead of stimulating phytoplankton blooms. These findings suggest that if large shellfish populations could be restored, these beds could play an important role in helping to achieve nitrogen reductions for mitigating the propagation of HABs across the globe.

Shellfish farms are delectable and nutritious nutrient sinks. Consider that a weekly harvest of 10,000 bivalves removes about 200 pounds of nitrogen per year; thus, a shellfish farm of about 1,000 acres would compensate for the nitrogenous wastes of about 50,000 coastal inhabitants.

Shellfish are a healthful addition to a balanced low-fat diet and are especially good sources of iron, zinc, copper, and vitamin B12. Furthermore, shellfish are much lower in saturated fat than pork or beef and provide high quality protein. Moreover, shellfish contain significant amounts of heart-healthy omega-3's, which are undetectable in chicken, beef and pork. Although salmon are one of the richest sources of omega-3 fatty acids at 1.17 grams per 100-gram edible portion, shellfish are close behind with mussels at .84 and oysters .6 grams per portion.

From a "harvest" perspective, sustainable shellfish aquaculture would provide greater global food security and when farmed locally, mitigate the carbon transportation footprint. From a "habitat" perspective, shellfish farms would provide enormous and valuable benefits to the global environment.

Molluscan Shellfish Aquaculture Worldwide

In the past five decades, global fisheries and aquaculture have grown steadily, and seafood consumption per capita has increased from an average of 9.9 kg in the 1960s to 19.2 kg in 2012. This development is impressive, and probably driven by multiple factors, such as population growth, rising income, and urbanization. Molluscan shellfish has traditionally been a major component of world aquaculture. For example, the molluscan shellfish production in 2012 (15.2 million metric tonnes; 1 metric ton equals to 1000 kilograms or 2204 pounds, hereinafter the same below) accounted for about 22.8% of the total (inland and marine) aquaculture production and 60.3% of the world marine aquaculture production.

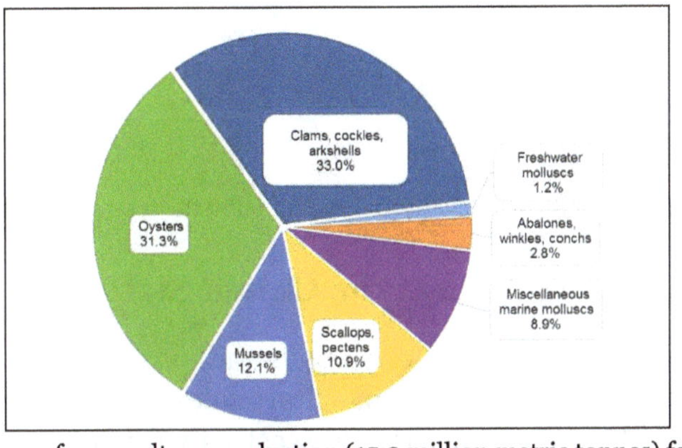

The percentage of aquaculture production (15.2 million metric tonnes) for the major molluscan species groups in 2012 worldwide

Worldwide, the major aquaculture molluscan species groups include clams (together with cockles and arkshells), oysters, mussels, scallops, abalones, and other miscellaneous molluscs. Based on their production in 2012, clams, oysters, mussels, and scallops were the four major groups and accounted for 87% of the total molluscan aquaculture production.

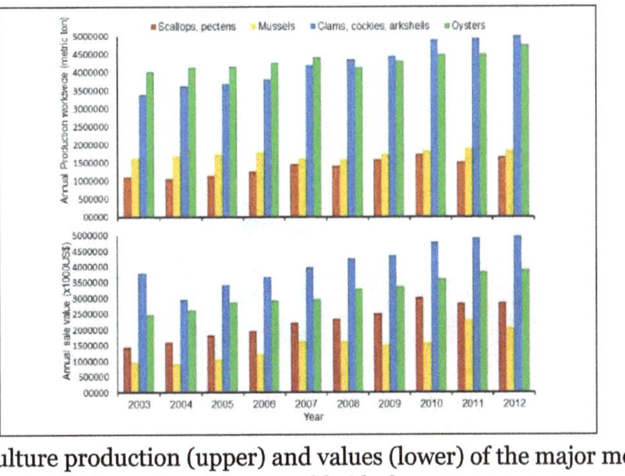

The aquaculture production (upper) and values (lower) of the major molluscan shellfish species groups worldwide from 2003 to 2012

The annual production (total weight) and sales values of the four primary molluscan aquaculture groups varied from year to year in the past decade. Oysters led annual production between 2003 and 2007 but have been replaced by clams since 2008. Clams represent the leading product in sales value, at about 5.0 billion USD in 2012, followed by oysters, scallops, and mussels.

Molluscs are cultured in 76 countries worldwide. The top ten countries for aquaculture production are China (81.4%), followed by Vietnam (2.6%), Korea Republic (2.5%), Japan (2.3%), Chile (1.7%), Spain (1.4%), Thailand (1.4%), USA (1.1%), France (1.1%), and Italy (0.7%). The major aquaculture species or groups differ by country.

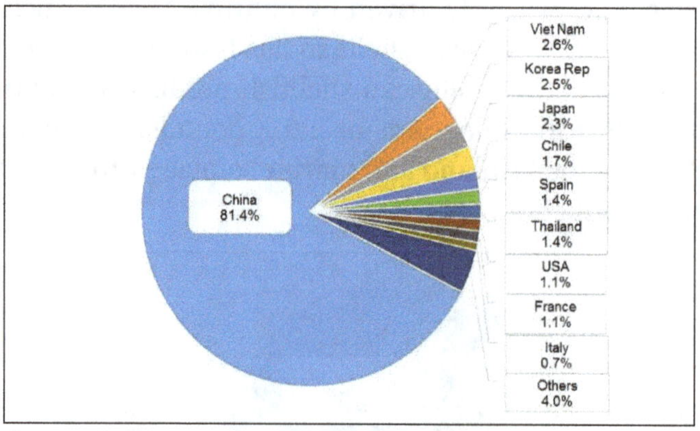

The percentage of molluscan shellfish aquaculture production in the top ten countries worldwide in 2012.

Molluscan Shellfish Aquaculture in the United States

In the United States, molluscan aquaculture is a $329 million industry involving 756 farms in 18 states. Nationwide, the top three species in aquaculture sales in 2013 were the Pacific oyster Crassostrea gigas ($87million), the eastern oyster Crassostrea virginica ($68 million), and the hard clam Mercenaria mercenaria ($65 million). The Pacific oyster is cultured on the west coast of the United States in the states of Alaska, Washington, Oregon, and California. The eastern oyster is cultured on the east coast of the United States and in the Gulf of Mexico in the states of Maine, Massachusetts, New York, Connecticut, Rhode Island, New Jersey, Maryland, Virginia, North Carolina, South Carolina, Georgia, Florida, Alabama, and Louisiana. The hard clam is also cultured on the east coast in the states of Massachusetts, New York, Connecticut, Rhode Island, New Jersey, Maryland, Virginia, North Carolina, South Carolina, Georgia, and Florida.

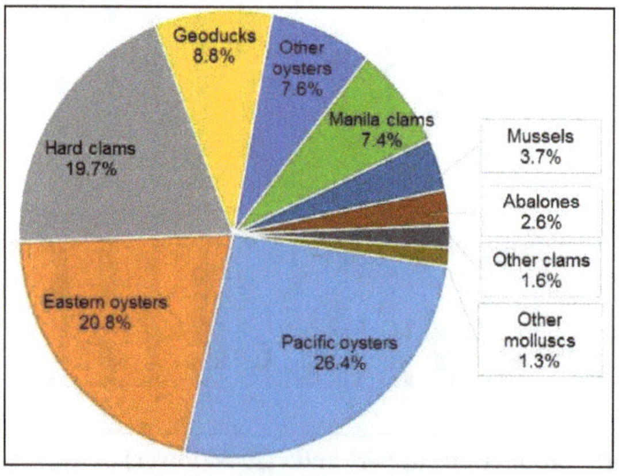

The proportion of aquaculture sales value for the major molluscan species groups in the United States in 2013

Among the 18 states with molluscan aquaculture, the state of Washington has the greatest aquaculture sales, at $149.3 million, or 45.4% of the total US sales. The molluscan aquaculture industry in the state of Washington is also the most diversified and includes Pacific oysters, mussels, manila clams, geoducks, and other molluscs. The states of Virginia ($41.5 million), Connecticut ($28.3 million), Florida ($19.6 million), and California ($17.0 million) trail Washington in molluscan aquaculture sales. In Virginia and Connecticut, the major species are eastern oysters and hard clams; in Florida, the major aquaculture species is the hard clam with some production of eastern oysters and sunray venus clams; and, in California, the aquaculture species are Pacific oysters, abalones, and clams.

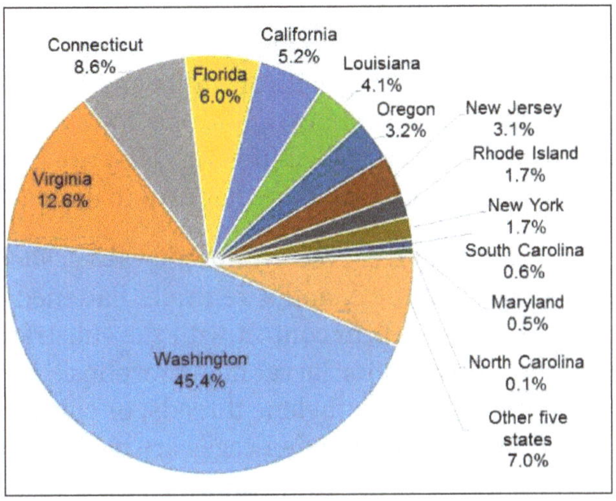

The allocation of molluscan aquaculture sales in the 18 states in the United States where molluscan shellfish are cultured in 2013

Molluscan Shellfish Habitats and Aquaculture

Most cultured molluscan species are bivalves; only a few gastropods, such as abalones, are cultured. Feeding and reproductive modes may account for the relative ease of culturing bivalves over gastropods. Bivalve molluscs are typically filter feeders of phytoplankton and are primarily free spawning with external fertilization. Due to the very large and diversified taxonomic class of gastropods, their reproduction, feeding, and habitat types vary greatly. Gastropod molluscs may be herbivores, carnivores, parasites, or omnivores, and exhibit complex mating behaviors (courtship) with either internal or external fertilization. Abalones are the primary gastropod species for aquaculture. Like bivalves, they are free spawning, with external fertilization, and have a free-swimming larval stage that feeds on phytoplankton. Following metamorphosis (a morphological transformation from swimming larvae to juvenile) the herbivorous spat (juveniles after metamorphosis) graze on benthic encrusting algae or bacterial films using their radula (an anatomical structure used in feeding that is unique to molluscs and present in all of them except bivalves). As abalone grow, they increasingly rely on macro algae as food sources; aquaculture of abalones requires providing microalgae-coated plates during larval metamorphosis and seaweed during growout (the culture period from seed to market-sized adults).

In general, bivalves for aquaculture typically have one of three habitat lifestyles: 1) endobenthic burrowing (bivalves that live within the sediments), 2) epibenthic cemented (bivalves that live on the surface of sediments), or 3) epibenthic free-living (non-attached) or attached with byssal thread (a keratin and protein filament secreted from the muscular foot). Most clams have an endobenthic burrowing lifestyle. The muscular, hatchet-shaped foot and round or elongated shell shape of most species of clams aid them in burrowing. Depending on the species, the preferred benthic substrate may be sand, mud, or a mixture, and burrowing depth may be from centimeters to meters. Successful aquaculture of these species requires suitable substrates for growout in open water environments. Oysters exhibit an epibenthic cemented lifestyle. When ready to metamorphose, free-swimming oyster larvae seek a suitable hard substrate, such as calcareous shell, and cement themselves down with a glue-like protein; they will remain attached for life and lose their foot after metamorphosis. Oysters can close tightly, providing protection from predators, desiccation, and other harsh environmental conditions. Culture of oyster species requires consideration of suitable substrate types for metamorphosis. During growout, oysters are cultured in a variety of systems, which vary from on-bottom to off-bottom, suspended or floating, with or without predator protection. Air drying and other control methods must be used to prevent biofouling of culture gear and oysters. Scallops and mussels have an epibenthic free or byssal attached lifestyle. They are capable of abandoning their byssal threads temporarily and swimming to avoid predators or to move to a more suitable location.

Scallops usually have two light-weight and curved symmetric valves, shell ears (the extended part of shell near the hinge) with a byssus orifice, and a ventrally flattened shape to enable them to create a directed water flow and ensure a firm contact with the substrate. Culture of scallop species requires providing suitable substrates for larval metamorphosis in the hatchery; commonly used substrates include window screen, polyethylene threads, or coir (natural fibre from coconut or palm tree) threads/ropes. Floating or suspended cages are used for scallop growout. Mussels can strongly attach to substrates by many byssal threads. Most mussels have two symmetric and thin shells hinged together tightly. Culture of mussel species also requires suitable substrates for larval metamorphosis in the hatchery. Growout culture usually occurs on ropes in floating rafts or off-bottom poles without cages due to their strong byssal attachment.

Molluscan Shellfish Aquaculture Stages and Methodologies

Molluscan shellfish aquaculture includes three biological or cultural stages: production of seed, juvenile nursery culture, and growout of sub-adults to harvest size. The aquaculture methods, culture gear, and routine maintenance must be based on a thorough understanding of the biological characteristics of the target species.

Seed Production

Two approaches are currently employed for seed production of molluscs: hatchery production and seed collection (collecting seed from natural resources by providing suitable substrates). In hatchery production, seed are produced under controlled conditions. Methods include broodstock selection and conditioning (to accelerate sexual maturity), spawning and gamete collection, fertilization, swimming larval culture, metamorphosis, and post-metamorphosis rearing. The water for hatcheries is usually treated by sedimentation, filtration, sterilization or disinfection to meet the needs of both the larvae and their multi-species microalgal food. Hatcheries have the advantage of increasing the quantity and quality of seed production, producing seed out of the natural spawning season, and providing opportunities for broodstock breeding and various genetic manipulations. In addition, hatchery-produced oyster pediveliger larvae (i.e., pre-settlement larvae) can be shipped to other locations for setting and further nursing, which is called remote setting. If allowed to settle on shell (i.e., cultch), the resulting seed, referred to as "spat-on-shell," may be used in extensive oyster aquaculture efforts, as well as restoration efforts. This method is in contrast to single oyster seed production, referred to as cultchless, which results in intensive culture of individual oysters in various containers (e.g., bags, cages).

The practice of collecting seed from the wild is usually performed by providing suitable substrates for metamorphosing larvae, such as rock piles, bamboo poles, and shell strings, or by tilling and cleaning natural substrates. Site and substrate selection, as well as timing, is critical for obtaining seed and minimizing labor cost. These methods have been used for oyster, mussel, and clam aquaculture for over two thousand years. Placing shell substrate, or cultch, during the natural spawning season is still a viable method of collecting oyster seed and supports extensive aquaculture efforts in many coastal states.

Nursery

The nursery stage is an intermediate stage between the hatchery and growout aimed to increase seed size in a protected environment prior to out-planting, thereby improving survival and growth. Based on the habitat requirements of the aquaculture species, nursery methods may include land-based flow-through systems (up- or down-wellers, raceways), floating or submerged nets or trays in intertidal ponds with enhanced microalgae culture, floating or bottom nets, or trays in open waters. Optimal seed density in the nursery varies depending on the species, seed size, food availability, and water flow. Routine maintenance during the nursery stage includes cleaning the culture systems, measuring water quality, and monitoring for disease.

Growout

After the nursery stage, larger seed or juveniles are out-planted and grown to harvest size during the growout stage of aquaculture. The first step is site selection; growout may take place in open waters, offshore, in coastal ponds, or in intertidal areas. All growout operations, whether located on private land or on public lands leased from a government entity must follow the federal and state regulations addressing shellfish harvesting water classifications, environmental issues, use of waterways, public health, and food safety, etc. In addition, understanding of the culture environmental conditions, such as wave and tidal height, water depth, substrate, salinity, temperature, phytoplankton biomass, and predator species and abundance, is the basis for culture method selection, layout, and management. Growout culture methods are species-specific and include on-bottom, off-bottom poles or racks, floating rafts or longline systems, and cages or nets of various designs. The selection of growout culture gear must consider the biological characteristics of the cultured species, the physical, chemical and biological characteristics of the culture environment, and the ease of routine maintenance and harvest.

Environmental Benefits of Shellfish Aquaculture

- Shellfish aquaculture is a sustainable and green industry.
- Shellfish harvest helps to improve sediment quality by loosening and dispersing silt and muck, and helps add oxygen to bottom waters and sediments.

- Oysters can filter 100 gallons of seawater in a single day as part of their natural feeding process. Shellfish feed on phytoplankton, but can also accumulate marine biotoxins, chemical contaminants, and pathogenic microorganisms, such as bacteria and viruses, effectively removing them from the water column.
- Shellfish improve water quality and clarity by removing particulates, excess nutrients, organic material, viruses, and bacteria from the water column. Improved water clarity enhances habitat for sea grasses such as eelgrass and other submerged aquatic vegetation.
- Shellfish help control harmful algal blooms, like Red Tide, by removing algal cells before they accumulate to harmful levels.
- Shellfish beds provide critical ecosystem functions by creating structure and habitat for other species such as crabs, worms, and juvenile fish, that provide a food source for fish and other marine species. The nooks and crannies in oyster beds create 50 times the surface area of an equal expanse of flat bottom.
- Shellfish beds stabilize sediments helping to protect the shoreline from erosion.
- Shellfish remove nitrogen from the environment in shellfish tissues that are removed when the animals are harvested.

References

- Fish-farming: awionline.org, Retrieved 14 June, 2019
- Fish-farming: roysfarm.com, Retrieved 27 January, 2019
- Fish-farming, environment: encyclopedia.com, Retrieved 7 July, 2019
- Mariculture, environment: encyclopedia.com, Retrieved 13 February, 2019
- Sustainable-shellfish-aquaculture: eco-business.com, Retrieved 23 August, 2019

Chapter 3

Aquaculture: Important Species, Diseases and Genetics

The welfare of organisms in aquaculture can be affected by a range of reasons such as disease, stocking densities, parasitism and behavioural interactions. The topics elaborated in this chapter will help in gaining a better perspective about the different species which are considered important in aquaculture and the diseases which afflict them.

Common Carp

Common carp (Cyprinus carpio) belongs to the order Cypriniformes and the family Cyprinidae, which is considered the largest family of freshwater fish. It generally inhabits freshwater environments, especially ponds, lakes and rivers, and also rarely inhabits brackish-water environments. It is widely distributed in almost all countries of the world but is very popular in Asia and some European countries. Because of its high popularity, its distribution has been widely extended by human introduction. Common carp is the third most frequently introduced species worldwide. It is being considered as a potential candidate for commercial aquaculture in Asia and some European countries as it has a very high adaptive capability to both environment and food. In some European countries, more than 80% of total fish production comes from common carp Common carp is frequently called an 'ecological engineer' because it can modify ecological characteristics of aquatic systems. In some western countries, it is frequently reported as a nuisance fish as it causes dramatic ecological disruption to both the aquatic community and ecosystem. In the USA, common carp is considered as the greatest threat to the biodiversity of wetland and shallow lake ecosystems. Therefore, many studies have been conducted to identify appropriate methods to control common carp populations in wetlands and shallow lakes.

Common carp is the third most widely cultivated and commercially important freshwater fish species in the world, it ranked third (grass carp ranked first and silver carp second) in terms of worldwide finfish aquaculture production contributing 9% of the world's total finfish aquaculture production, and Asia accounted for more than 90% of common carp's aquaculture production China alone contributed 77% (2,462,346 tons) of the world's aquaculture production of common carp (3,216,203 tons) in 2009. In Asia, common carp is normally cultured in various aquaculture systems but the most common is the semi-intensive pond polyculture system.

Common carp is a very well-known benthivorous fish that has larger bottom–up effects than other benthivorous fish. The bottom–up effects of common carp mainly depend on the incorporation of benthos-derived nutrients and the release of nutrients from bottom sediment during grazing on benthos. The productivity of freshwater systems is often limited by a lack of soluble phosphorus (PO_4 – P), which acts as a limiting nutrient for phytoplankton growth. Common carp enhances phytoplankton productivity by releasing nutrients including soluble phosphorus from the sediment. This accelerates the nutrient fluxes to the next trophic levels. This positively influences the production of fish, which depend directly or indirectly on natural food. For example, rohu (Labeo rohita) is a planktivorous fish, which grows better in ponds with common carp than in a monoculture. Therefore, common carp is commonly reared in polyculture ponds; however, in cold climate regions, especially in Europe, monoculture of common carp is more popular.

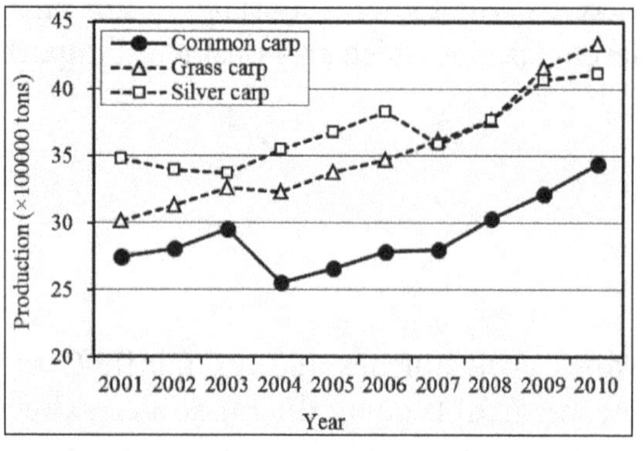

World aquaculture production trends of common carp, grass carp and silver carp.

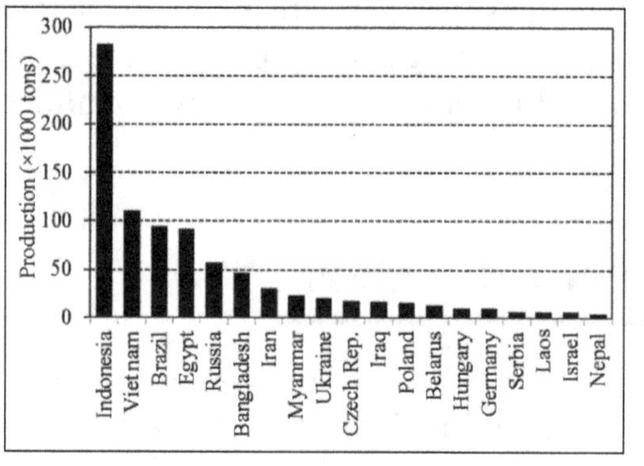

Major common carp-producing countries (except China) and their production in

The density of animals is a critical factor that strongly affects the aquatic ecosystem. Excess densities of common carp have many negative effects on aquatic ecosystems. For example, after reaching a critical density, common carp changes the physical condition of an aquatic system from clear-water conditions dominated by macrophytes to a turbid state with a loss of macrophytes. In polyculture ponds, the optimum density of common carp can improve synergistic effects with other fish, which potentially increase nutrient retention efficiency in fish and decrease nutrient loss in the sediment.

When food resources are not sufficient, common carp changes its food habits, feeding niche and behaviour. Common carp also influences other fish species to change their food habits and behaviour, which affects their growth and behaviour either positively or negatively.

Effects of Common Carp on the Aquatic Environment

The nutrient turnover of the aquatic environment is strongly influenced by the fish population, especially planktivorous and benthivorous fish populations. The majority of nutrients in ponds are stored in bottom sediments, in both organic and inorganic forms. Sediment can store more than 100 times more nutrients than the water column. The transfer of nutrients back into the water column by the resuspension of sediment can have an important influence on the limnology of ponds. Fish species, size, density, food availability and foraging behaviour (benthic feeding) are important critical factors that greatly affect sediment resuspension. Several species often resuspend bottom sediment; the best known is the common carp, which has strong effects on the aquatic environment owing to its browsing activity for benthic macroinvertebrates in the sediment.

Common carp is specialized to browse for benthic macroinvertebrates in the sediment. By doing this, it affects water visibility by resuspending clay particles in the water, cycling of nutrients, and affecting the abundance of phytoplankton, zooplankton and benthic macroinvertebrates. Among various benthic macroinvertebrates, chironomid larvae are an important benthic food source for common carp. Depending on species, size and type of sediment, chironomid larvae live up to several centimetres deep in the sediment. Therefore, common carp requires a special technique to pick up chironomids from sediments. The carp separates benthic macroinvertebrates by digging and sieving of sediments. Its digging activity for browsing benthic macroinvertebrates is effective down to approximately 3 cm. During the digging and sieving of sediments, it causes bottom soil resuspension, which increases oxygen availability in the bottom soil. Generally, diffusion is an important mechanism to make oxygen available from the water column into the bottom soil, but this process is very slow and extremely insufficient, especially in natural feed-driven aquaculture systems. However, resuspension by common carp increases aerobic decomposition in the sediment by increasing oxygen availability in the bottom soil. This process accelerates the mineralization of organic matter in the bottom soil. The mineralization of organic matter happens more rapidly under aerobic than under anaerobic conditions. Therefore, favouring aerobic decomposition of organic matter stimulates nutrient cycling in ponds, which in turn has a large impact on the abiotic and biotic properties of the overlying water column Common carp accelerates the decomposition of organic matter in the following two ways:

1. Decomposition in the bottom soil: In general, oxygen is mixed in to the bottom soil by diffusion. Wind and mechanically driven water currents mix the water column and bring some oxygen to the bottom layers. However, in most cases, this process is not sufficient. Common carp improves the oxidizing conditions in the bottom soil by disturbing the transition zone between the pond bottom soil and the overlying water. The transition zone between the pond bottom soil and the overlying water varies greatly under dynamic conditions. The transition between oxygenated overlying water and anaerobic soil takes place along a gradient of a few millimetres. However, disturbance of the bottom soil by common carp raises the depth and extent of oxygen penetration into the soil. The accelerated aerobic conditions due to carp bioturbation leads to increased decomposition and mineralization of organic matter in the bottom soil.

2. Decomposition in the water column: In general, the upper layer of the pond bottom soil is enriched with organic matter, which cannot be decomposed entirely owing to a lack of oxygen. Resuspension of bottom soil by common carp induces decomposition by exporting the organic matter to the water column, where the concentration of oxygen is generally high.

Dissolved oxygen is reduced during aerobic decomposition. Carbon dioxide is also released during decomposition, and contributes to lower pH and reduced alkalinity. Resuspension increases the mineralization rate, with increased nitrate nitrogen (NO_3-N), total ammonia nitrogen (TAN), total nitrogen (TN) and phosphate phosphorus (PO_4-P) and total phosphorus (TP) in the water. Common carp also substantially accelerates nitrogen and phosphorus transport from the bottom sediment into the water column via excretion. The increased nutrients stimulate photosynthesis, thus increasing phytoplankton biomass in the water column. Hytoplankton production is limited by low PO_4-P concentration in most freshwater ponds and lakes. Resuspension driven by common carp decreases this limitation by increasing P flux from the sediment to the water column. Zooplankton production rates increase owing to the enhancement of phytoplankton biomass in the presence of common carp.

The density of common carp is a very critical factor for the aquatic environment. A substantial impact on aquatic ecology occurs after exceeding the critical common carp density, which largely depends on the habitat. For example, in shallow unfertile lakes, submerged macrophytes can be significantly affected by common carp when the common carp biomass approaches 200 kg ha^{-1} whereas a fertile lake can support up to 1125 kg ha−1, which is some five times higher than in shallow unfertile lakes. In the case of polyculture ponds, water quality, abundance of plankton and benthic macroinvertebrates, and fish growth can be strongly affected when the common carp density approaches more than about 1000 kg ha^{-1}.

Artificial feeding is another very important factor that influences the critical density of fish. The critical density can be doubled if pelleted artificial feed is supplied to common carp. In the presence of artificial feed, common carp shifts its preference from zooplankton and benthic macroinvertebrates to artificial feed. Common carp grows better in ponds supplied with pelleted artificial feed than with extruded feed and cereals. Pelleted artificial feed acts as a source of nutrients for common carp growth, but it also indirectly maintains ecological stability and controls cyanobacterial blooms in ponds. Therefore, common carp can be stocked at high density in polyculture ponds supplied with artificial feed without any negative effects on water quality and natural food resources. Rahman observed 1628 kg ha^{-1} common carp biomass in ponds supplied with artificial feed without any negative effects on water quality and natural food resources. However, an excess of common carp has several negative effects on pond or lake ecology.

1. Decreased soluble phosphorus in the water: Excessive resuspension of sediment may increase the redox potential. An increased redox potential may have a positive effect on the precipitation of soluble phosphorus (PO_4-P) through the formation of phosphate-rich inorganic particles, e.g. with iron as iron (III) phosphate. Sometimes, nutrients appear to be abundant in the water but the pond has a relatively low primary productivity of algae owing to a lack of bioavailable phosphorus (PO_4-P). Although this pattern may be linked to the complex chemical interactions in water, excess common carp decreases the concentration of (PO_4-P) in the water.
2. Increased turbidity of water: A high resuspension of bottom soil increases turbidity, which reduces light penetration. The biomass of planktivorous fish decreases with increasing turbidity and their resulting inability to locate and feed on plankton. High turbidity also affects the growth of submerged vegetation in shallow lakes and ponds. Common carp may directly consume many types of macrophytes and indirectly uproot or break macrophytes while searching for benthic macroinvertebrates. However, the carp's indirect effects are more severe than its direct effects on aquatic macrophytes. Common carp density has a strong negative linear correlation with vegetation yields. Areas macrophytes exist provide shelter and food for invertebrates and larval fish. Therefore, many invertebrate groups may be reduced in abundance and the invertebrate community composition may be altered if macrophytes are reduced by common carp activity. Biodiversity in the aquatic ecosystem can be severely affected by a high density of common carp.
3. Decreased system production: High turbidity and low phosphorus concentration reduce photosynthesis. Subsequently, primary, secondary and tertiary production in the aquatic

> ecosystem is reduced. At high density, common carp increase grazing pressure on natural food, especially zooplankton and benthic macroinvertebrates, to such an extent that recovery is not possible. Therefore, the abundance of zooplankton and benthic macroinvertebrates can be strongly suppressed at a high density of common carp. Depending on the environment, there is a threshold in the density of common carp above which zooplankton and benthic macroinvertebrate populations collapse.

Schematic representation of the effects of common carp on nutrients (PO_4^-P, PO_4^-P, NO_3^-N, NO_2^-N, total ammonia nitrogen) and natural food availability. + and – indicate positive and negative effects, respectively. Filled ovals indicate processes and states that illustrate the influence of common carp on primary production.

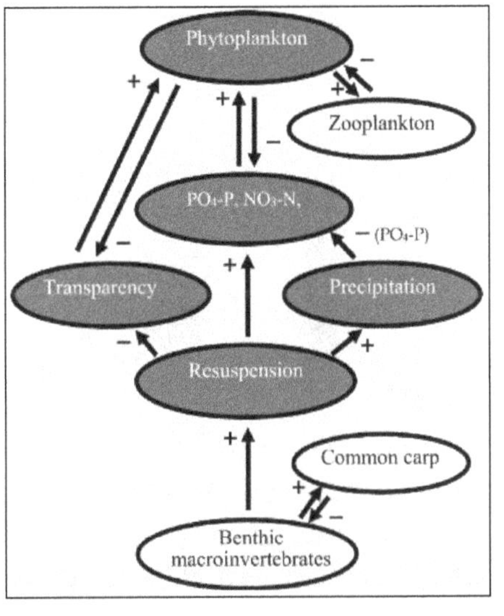

The effects of common carp on environmental conditions are also size dependent. Aquatic environments are differently affected by different ontogenetic stages of common carp. However, changes in environmental conditions are not linearly related to common carp size. The effect of common carp size on the aquatic environment almost disappears when the fish are more than 1 year old.

Effects of Common Carp on Fish Production

Resuspension of bottom soil by benthivorous fish affects not only the aquatic environment but also the level of fish production. Thus, increasing fish production by the addition of common carp is a common practice in many parts of the world, including Asia and Europe. This technique is very useful for semi-intensive polyculture systems where fish production depends almost entirely on natural food. The interspecific interactions among fish species are important in the sustenance of any polyculture system Polyculture systems increase the utilization of the natural foods in fish ponds, by stocking a proper combination of two or more fish species, especially planktivorous and benthivorous fish, at proper densities. This strongly influences the nutrient turnover of the pond. For example, when planktivorous fish are kept together with common carp, the ponds generally require 20–40% less fertilizer to maintain adequate natural food levels than those with planktivorous fish in monoculture. Nitrogen and phosphorus retention efficiency in fish biomass increases in polyculture with planktivorous fish and common carp. This results in more nutrients passing through the pond's food web, and fewer nutrients in the bottom soil.

Comparison of the effect of culture systems with common carp (CC) (polyculture of rohu and common carp) and without CC (monoculture of rohu) on nitrogen and phosphorus accumulation (%) in bottom soil and fish. Nitrogen and phosphorus retention efficiencies with no letters in common among three attached bars are significantly different ($p<0.05$). Retention efficiency is calculated based on total input nutrients, higher total nutrient input and total fish.

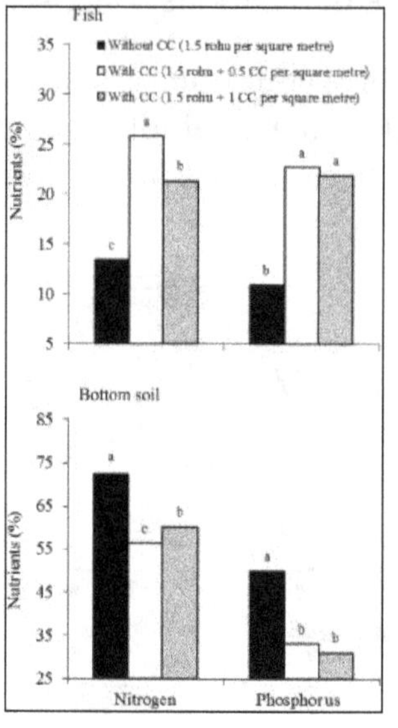

Many benthivorous carp species can be used in polyculture to increase the production of planktivorous fish. These include common carp mrigal. Among these, the best known and most widely used in semi-intensive polyculture is the common carp. The common carp has a greater effect than any other benthivorous fish on aquatic ecology and fish production bserved 1.6 times higher yield of rohu in the presence common carp than in the presence of mrigal in semi-intensive polyculture ponds. In another study, observed 1.5 times higher growth of rohu in the presence of common carp than in the presence of calbasu. Parkos et al. observed greater effects on phytoplankton and total phosphorus availability in the presence of common carp than in the presence of channel catfish (Ictalurus punctatus). Common carp (under optimum density) has a strong synergistic effect in terms of producing natural food, which is used as food for other fish living in the water column. Therefore, total fish production is increased in the presence of common carp. Common carp also influence aquaculture production indirectly by creating favourable conditions at the pond bottom, or by increasing the diffusion of toxic compounds (e.g. organic acids, manganese and sulphur compounds) from the bottom soil upwards to the water column.

Table: Growth and production of fish in systems with various densities of common carp.

Variable	System	Common carp density	Average value
SGR of rohu (% day−1)	Pond	0	1.39
	Pond	0.5 m^{-2}	1.55
	Pond	1 m^{-2}	1.35
	Simulated pond	0	0.9
	Simulated pond	1 m^{-2}	1.2
	Simulated pond	2 m^{-2}	1.1

Rohu production (kg ha−1 137 days−1)	Pond	0	1747
	Pond	0.5 m^{-2}	2473
	Pond	1 m^{-2}	1716
Total fish production (kg ha−1 137 days−1)	Pond	0	1747
	Pond	0.5 m^{-2}	3532
	Pond	1 m^{-2}	3030
Total fish production (kg ha−1 135 days−1)	Pond	0.5 m^{-2}	2123

Stocking density is important not only for common carp but also for planktivorous fish to obtain high production levels from a polyculture system. Pond productivity or fertility is a very important factor in calculating the optimum stocking density in a polyculture system. The stocking ratios vary widely in different areas, based on the management techniques used. If the stocking ratio of benthivorous and planktivorous fish is not appropriate, then many negative effects on fish growth and production can be observed. For example, rohu growth can be lower in ponds with a higher density of common carp (1 common carp m^{-2}) than in ponds with lower stocking density (0.5 common carp m^{-2}). A similar effect of common carp density on the growth of rohu was also observed in tanks simulating the pond environment. In large lakes, common carp biomass is often negatively related to the abundance of other fish, especially bluegill (Lepomis macrochirus), black crappie (Pomoxis nigromaculatus), largemouth bass. (Micropterus salmoides), smallmouth bass (Micropterus dolomieu), black bullhead (Ameiurus melas), walleye (Sander vitreus), yellow perch (Perca flavescens), northern pike (Esox lucius) and white bass. The most important reason may be habitat degradation by common carp, which increases the turbidity of the water in the lake through its benthic foraging behaviour, switching lakes from the clear- to the turbid-water state.

Effects of Common Carp on Food Habits, Feeding Niche and Behaviour

Food habits of fish are highly variable and depend on a wide variety of factors, including the species and age of the fish, the availability of preferred food and the combination of fish species. Fish consume different food organisms in different amounts depending on the availability of the food items and the combinations and densities of species. The proper combination of fish species may result in synergism. For example, strong synergistic effects in terms of food availability, food intake, growth and production of fish can be obtained in rohu ponds with 0.5 common carp per m^2. These effects nearly disappear in rohu ponds with 1 common carp per m^2. The correct stocking density influences individual food availability, and high densities cause preferred foods to become depleted. This leads common carp to shift its food habits. Common carp preferentially consumes benthic macroinvertebrates. A low abundance of benthic macroinvertebrates or a high stocking density of common carp could cause the carp to switch to its next preferred food items (zooplankton), leading to significant dietary competition with other fish which prefer zooplankton. However, common carp has excellent adaptive capabilities in the presence of insufficient food. There is evidence that common carp eats the fry of other fish at high density, when there are insufficient other natural foods. There is also evidence that the common carp predates on crayfish (Cambarellus montezumae) larvae when they live in the same habitat.

Common carp has mechanisms to maximize fitness not only by shifting food preference but also by modifying its feeding niche and behaviour. Common carp can modify its feeding niche and behaviour in the presence of superior fish, making it a valuable species not only for monoculture but also for polyculture. When two species forage on the same limited food resources, interspecific competition for food will be intense. In this situation, common carp follows the classical optimal foraging theory, whereby it broadens its feeding niche to maximize its food intake.

Common carp is an omnivorous fish that primarily feeds on benthic macroinvertebrates (chironomids, tubificids) and zooplankton, but the bulk of its diet consists of detritus. Therefore, the feeding niche of common carp in natural systems is largely benthic. Common carp generally ignores phytoplankton, strongly selects benthic macroinvertebrates and weakly selects zooplankton. These results suggest that common carp prefers benthic macroinvertebrates to zooplankton when plankton and benthic macroinvertebrates are provided together. This preference for benthic macroinvertebrates probably influences the behaviour of the common carp, as its spends the majority of its grazing and swimming time near the bottom in an aquatic system. When preferred food (benthic macroinvertebrates) becomes depleted, common carp increases its preference for zooplankton, which can be a very dominant food in the absence of benthic macroinvertebrates. The dependency on zooplankton shifts the common carp's feeding niche from the benthic zone to the water column. In this case, common carp spends the majority of its grazing and swimming time in the water column.

The presence of common carp can also alter the behaviour of other fish. Rohu is a planktivorous fish, which spends the majority of its grazing and swimming time in the water column. Rohu grazes more than three times longer in the water column than near the pond bottom. In the presence of common carp, rohu decreases its grazing time in the water column and increases its grazing time near the pond bottom. This effect is greater at higher than at lower densities of common carp. Rohu also increases its time spent actively swimming and searching for food in the presence of common carp. Common carp influences the behaviour not only of fish but also of crustaceans. Crayfish generally lives in a habitat where submerged macrophytes (e.g. Potamogeton pectinatus) and algae (e.g. Cladophora glomerata) are available. When submerged macrophytes and algae are affected by the presence of common carp, crayfish not only displaces its habitat but also increases its displacement speed.

Ontogenetic Shifts in Dietary Preferences of Common Carp

Ontogenetic shifts in diet are very common in fish. Many factors are responsible for these changes in diet, and can be divided into two categories: external factors (e.g. habitat, food supply, predation risk) and internal factors (e.g. anatomical structures, behaviour, physiological demands). In many species, dietary changes are associated with habitat shifts. Changes in the size of the mouth and oral anatomy may also correspond to ontogenetic dietary shifts. Like many fish, common carp shows ontogenetic shifts in food habits.

Common carp is a benthivorous fish feeding mainly on benthic macroinvertebrates. Although this statement is commonly accepted, it is not true for all common carp. Small common carp are known to feed preferentially on zooplankton, whereas larger common carp avoid zooplankton and concentrate on benthic macroinvertebrates. common carp of up to 15.4 cm total length preferentially select zooplankton, but common carp larger than 18.9 cm total length avoid this food item. Ingestion of benthic macroinvertebrates by common carp increases significantly with its increasing size, but common carp size has no significant effect on phytoplankton consumption. All sizes of common carp avoid phytoplankton. Overall, the proportion of zooplankton ingestion decreases with increasing common carp size. The diet of common carp is influenced by many factors, including the age of the fish, the availability of natural food and the season.

Effect of Common Carp on the Diel Feeding Rhythm

Diel activities of most fish species are mainly synchronized with alterations in day and night. Fish can be classified as either diurnal feeders which rely on vision, or nocturnal feeders which rely more on tactile, chemical or electrical senses. However, diel activities are largely species specific. Some fish search for food during both dark and light periods, but are more active during daytime. The grazing behaviour of those fish depends on both light and the availability of food. Common carp is a very active fish, which grazes during both day and night but prefers to graze during daytime. It shows the opposite pattern for non-feeding swimming behavior Diel variations in the vertical swimming behaviour of common carp are related to available food types. In the absence of benthic macroinvertebrates, it spends most of its time in the water column, whereas when benthic macroinvertebrates are present, it spends more time near the pond bottom.

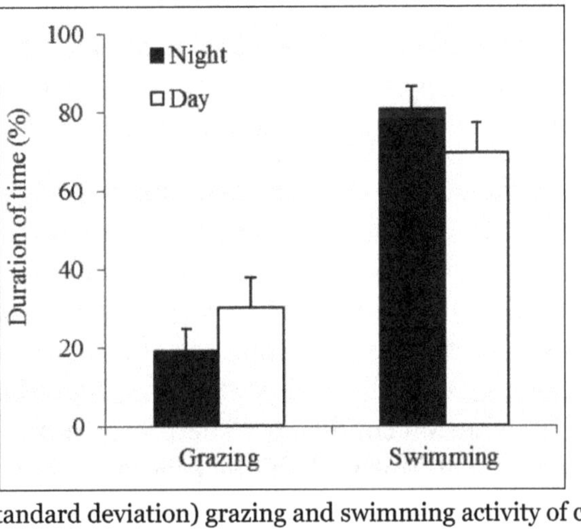

Mean (bar: ± standard deviation) grazing and swimming activity of common carp.

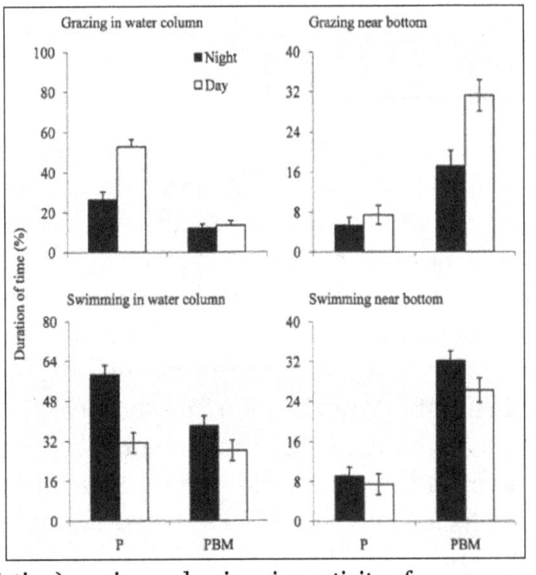

Mean (bar: ± standard deviation) grazing and swimming activity of common carp in simulated ponds with only plankton (P) and simulated ponds with plankton and benthic macroinvertebrates (PBM).

Common carp affects the diel feeding rhythms of other fish. For example, the presence and density of common carp affect the diel rhythms of rohu in polyculture ponds. Rohu is a diurnal feeder, spending the majority of its grazing and swimming time in the water column during daytime. In the presence of common carp, the difference between day and night grazing and swimming in the water column by rohu decreases, and the difference between day and night grazing and swimming near the pond bottom increases. This effect is greater at higher than at lower densities of common carp.

In conclusion, aquaculture of high-value carnivorous fish is rapidly expanding to bridge the gap between supply and demand. Fish from natural food-driven aquaculture receive low priority because of their low demand and slow growth rate, and because they are difficult to culture. However, carnivorous fish consume considerably more fish protein than they produce. culturing of carnivorous fish uses up to five times more fish protein than is produced. The protein that is not retained by cultured fish causes various environmental problems including eutrophication and disease outbreaks. The major source of aquafeed protein is fishmeal produced from trash fish/low-value marine fish. Therefore, some marine fish stocks may be decreasing with the rapid expansion of carnivorous fish farming. According to the around 7% of marine fish stocks have been totally depleted, 17% are over-exploited and 52% are completely exploited. The expansion of natural food-driven aquaculture with benthivorous and filter-feeding fish may minimize this problem. Filter-feeding fish species depend on natural productivity, which may be enhanced by stocking benthivorous fish such as common carp.

Common carp can be used as a management tool to manipulate the system ecology to achieve high growth and production of filter-feeding fish. However, more research is needed to understand the complex interactions between common carp and locally available filter-feeding fish. Although many studies have focused on the effects of common carp on pond and lake ecology, it is still difficult to separate the biotic and abiotic processes that are affected by common carp because biotic and abiotic processes are strongly interrelated. For example, depredation of phytoplankton grazers (filter-feeding fish) reinforces turbid conditions. Therefore, more research is needed to understand the effects of common carp on biotic and abiotic processes separately.

Diseases of Common Carp

The widespread distribution and adaptability of the common carp, not to mention the attention resulting from its commercial importance, mean that a vast number of associated parasitic species have been identified. The most comprehensive list compiled so far includes 310 parasite species, including representatives of all major taxonomic phyla ranging from myxozoa to arachnidae. Like other host organisms, individual carp may be infected by a whole range of parasitic species, known collectively as the parasite infra community. Guegan and Kennedy have found a clear relationship between the length of time a host species has been present in a region and the species richness of its associated community of helminth parasites. For common carp, therefore, we can expect the parasite community be richer in its native Asian range than in parts of its introduced range such as Europe, where it was brought in the mediaeval period. Instead of reiterating a list of known parasites for common carp, the focus here will be on selected examples from all major groups, including a description of the most important associated diseases in feral and farmed fish and, where available, suggestions for prevention and cure.

Viral Diseases

The majority of viral fish diseases recognized today have been identified within the last few decades, not because they are new, but because developments in molecular diagnostics have revolutionized the study of fish diseases in the last 50 years. New tools such as the amplification of nucleid acids via Polymerase Chain Reaction (PCR) and subsequent sequencing of informative genes, restriction enzyme digestion and probe hybridization have facilitated research into many previously mysterious viral fish diseases worldwide.

Rhabdoviruses

Of special importance among carp viruses are Rhabdovirus carpio, the agent of Spring Viraemia of Carp (SVC) and certain fish herpesviruses, which result in significant economic losses in aquaculture and increasingly also in wild fish. SVC can occur if spring temperature rises above 7 °C, causing serious mortality in carp of all ages. Maximum mortality occurs between 10–15 °C but the virus may survive at temperatures up to 23 °C. It can infect all varieties of Cyprinus carpio (mirror-, leather-, koi-, grass-, silver-, bighead- and crucian carp and goldfish) as well as other species including pike (Esox Lucius), tench (Tinca tinca), roach (Rutilus rutilus) and wels catfish (Siluris glanis). Clinical signs of disease are somewhat unspecific and can be further confused by accompanying bacterial infections; they include lethargy, dark body coloration, decreased ventilation rate, exophthalmia (pop-eye), dropsy and bleeding from gills, fins and internal organs. SVC is transmitted via feces but it has been shown that blood sucking parasites such as leeches (Piscicola geometra) and the carp louse (Argulus foliaceus) can act as vectors for transmission to uninfected fish. Suspected cases of SVC should be properly diagnosed by an appropriate laboratory. Unambiguous identification can be achieved either by direct PCR-based methods such as sequencing of certain genes or by indirect methods such as ELISA, which uses antibodies and a color change as a diagnostic tool to detect an infecting agent. Once the disease is properly diagnosed, control is best achieved by maintaining the water temperature above 20 °C. Currently no antiviral drugs are available to treat SVC, but recent research has found promising results in the development of a vaccine.

Herpesviruses

The herpesviruses are a group of highly host-specific large DNA viruses, of which three recognized taxa are known to infect common carp including koi and goldfish. The carp pox herpesvirus CyHV-1 causes a disease known as carp pox (CHV), which typically causes white or gray ulcers in the epithelium of common carp and koi. Cyprinid herpesvirus 2(CyHV-2) causes hematopoietic necrosis in goldfish, and koi herpesvirus CyHV-3 (KHV), affects carp and its varieties including koi and ghost carp.

KHV, was first reported in 1998 by Ariav et al., but is now endemic in Eurasia, Africa and North America, where it is responsible for enormous economic losses in carp industries. Problems with KHV are increasingly reported from wild carp populations where infection can cause mass mortality events such as recently documented in Canada. Outbreaks of KHV occur seasonally when the water temperature rises above 13 °C. At lower temperatures, the virus remains dormant in experimentally infected fish, but causes rapid mortality within 7–12 days following a temperature shift to 23 °C. Afflicted fish are lethargic and exhibit pale patches on the skin. The gills initially appear pale, with irregular coloration and eventually exhibit epithelium necrosis. Eyeballs are sunken (enopthalmus) and mucus production is increased. Fry and juveniles are also affected and infection can result in high rates of mortality. The virus is transmitted horizontally, via viral particles excreted into the water along with feces. There is evidence that the virus can persist as a latent infection in the host and thus be spread even by seemingly healthy fish, then activated by temperature stress. As with SVC, the most efficient methods for diagnosing KHV are based on PCR or ELISA testing.

The severe threat posed by KHV to global carp production has resulted in intensive research and the development of efficient methodologies for prevention and treatment have identified potentially KHV-resistant local carp lineages and tried to select for this trait in crossbreeding experiments. Ronen et al. took an induced resistance approach, exposing carp to the virus for 3–5 days at 23 °C, then increasing the water temperature to 30 °C. Following this procedure, high levels of specific antibodies were found in the sera of carp, indicating resistance to KHV. Perelberg et al. found that carp that survive at least two natural outbreaks of KHV became resistant to the disease, and went on to develop an inactivated vaccine. However the protective effects were short-lived and treated fish required regular repeat immunizations. The newest approach is the development of a promising DNA vaccine by Zhou et al. which may offer long-term protection against KHV.

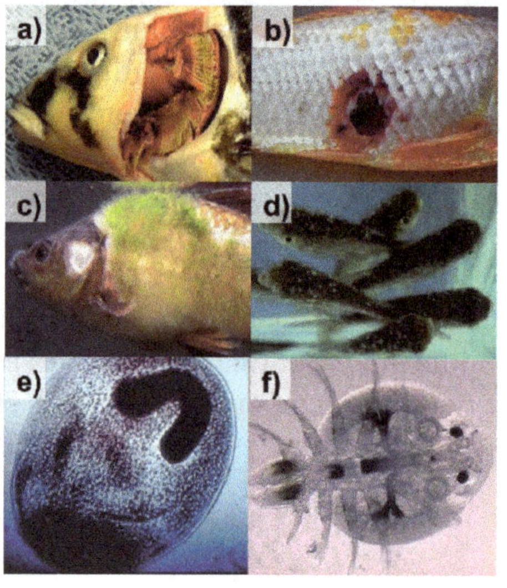

In figure above shows, a. Koi infected with KHV showing irregular gill colour, necrotic tissue and enopththalmus; b. Ulcer on fish body caused by an Aeromonas sp. infection; c. Carp with Aeromonas sp. Infection d. grass carp fingerlings with white spot, Ichthyophtiriusmultifilis; e. microscopic image of Ichthyophtiriusmultifilis, showing the horseshoe-shaped macronucleus; f. The carp louse Argulusfoliaceus.

Bacterial Diseases

Vast numbers of bacteria are associated with fish diseases, but, not all of the species involved are primarily fish pathogens. Many are opportunistic pathogens that only infect the host when it is already weakened or immunologically compromised. Most bacterial diseases typically appear as secondary infections in fish that are already stressed or carrying other infections. As with viral diseases, research into bacterial disease has greatly benefited from recent technical developments in molecular genetics.

Flexibacter

One of the most widespread bacterial diseases in both wild and farmed fish is columnaris disease, caused by Flavobacterium columnare. First described in 1922 by Davis in fishes from the Mississippi River, the bacterium can infect a wide range of freshwater fishes and causes severe losses in fish farms worldwide. F. columnare is a component of the bacterial micro biota of freshwater fish, fish eggs and water. It can survive and remain infective for several months outside its host, in pond sediment or even in sterile water.

F. columnare occur as both acute and chronic infections, typically affecting the gills, skin and fins of the host. The severity of infection and the progress of the disease depend on the virulence of the bacterial strain involved, and this may vary considerably. Only fish infected with milder strains show external signs of disease, as the more virulent strains tend to be fatal within a few days, before any external symptoms of columnaris become apparent. The first signs of chronic columnaris are small lesions on the fish's body, starting as pale discolorations of the skin. These are typically surrounded by a reddish zone. These usually first appear at the proximal basis of the dorsal fin, and progress to the distal edge, contrasting with 'normal' fin rot in which the damage starts at the outer edge of the fin and progresses towards the base. On the fish's body, the discoloration caused by columnaris progresses from the fin base in a saddle-like manner, hence the alternative name of 'saddle-back' disease. As the disease advances, lesions develop on the head and tail and penetrate deeper into the skin and muscle tissue layers resulting in deep ulcers typically covered with white-yellowish mucus. Often secondary infections with other bacteria and/or fungi aggravate the condition. Discoloration can also be seen on the gills, where yellowish-white-areas of degeneration indicate necrosis, which may cause respiratory distress.

As with viral diseases, bacterial infections such as columnaris are best diagnosed using PCR methods. The advantage of PCR lies in the ability to detect F. columnare at very low levels, in tissues such as gills or skin, thereby allowing to prevention and/or control measures to be initiated early. As with many other diseases, the best prevention is to reduce the stocking density in aquaculture facilities or rearing ponds, especially during periods of elevated water temperature. Declercq et al. suggest that since reducing rearing densities may also provide fewer opportunities for the transmission of ectoparasites and penetrating endoparasites, it may be an efficient general tool in the ecological management of fish disease.

Other prevention methods have been described, including treatment with oral antibiotics such as oxytetracycline, which can be administered along with the feed, and vaccination with bacteria killed by exposure to heat or formalin. Orally delivered oxytetracycline has proven an effective cure in both the early and advanced stages of disease outbreaks. A very interesting new approach to columnaris treatment may be the use of F. columnare-specific phage lysates. Prasad et al. reported 100% survival of phage-treated walking catfish Clarius batrachus in experimental infection trials.

Aeromonas

Several bacteria from the Aeromonas species complex are well known pathogens of fish, and many strains new to science have been identified in the last 10 years. Aeromonas species normally occur in fresh, brackish and coastal waters and as opportunistic members of the intestinal micro flora of many fish species, where they are usually tolerated by healthy individuals. However, in immunocompromised hosts, secondary infection with Aeromonas can be pathogenic. The best studied species are Aeromonas salmonicida, A. hydrophila and A. sobria, and all three have the potential to cause sub-acute to chronic disease in carp and other cyprinids. A. salmonicida, a non-motile aeromonad, is the cause of carp erythrodermatitis. In the course of the disease, roundish ulcers form as a result of epithelial necrosis, and fish may lose their appetite. Mortality is normally low. A. hydrophila and A. sobria are motile aeromonads. Infections occur mainly in spring and summer when the temperatures rise. Fish showing signs of A. hydrophila or A. sobria infection are often either co-infected with other pathogenic bacteria, suffering from injuries or living under stressed conditions such as intensive aquacultural facilities. Symptoms of acute A. hydrophila or A. sobria infection appear as visible damage on the body surface, such as fin or tail rot, large ulcers and skin damage, and bleeding at the base of fins, the anal vent and the gills. Systemic infection results in severe symptoms including damage to all organs, dropsy and death. Although motile aeromonads can be found almost everywhere, they are much more abundant in polluted water and especially in water with a high organic load, as is often found in densely stocked aquaculture ponds. Under such conditions, fish are continually exposed to high numbers of bacteria. Low oxygen levels, high fish densities, high temperatures and other stress factors increase the likelihood of annual outbreaks of the disease, mostly in spring. Below 4 °C, Aeromonas spp. are thought to be inactive. However activity and infection rates rise exponentially with temperature increases. At temperatures over 12 °C, the fish begin to build immunity to infection, but under stressed conditions this immune response may fail. The best protection against Aeromonas infections within cultured carp populations is thus a reduction of stress. As for columnaris, the recommended treatment is antibiotics delivered in fish feed . Loss of appetite in sick fish may limit the uptake of the antibiotic medication and any benefit should be weighed against the risk of antibiotic resistance developing in bacteria.

Fungal Diseases

Fungal infections can lead to significant losses in both free-living and cultured fish. Most fungal diseases of fish appear as secondary infections in hosts with pre-existing injuries or bacterial infections. The fungi can enter the body via wounds or ulcers (such as those caused by bacteria like Aeromona ssp. described above) and proliferate quickly, often either causing or significantly hastening death. The major pathogenic fungi (known as water molds) of concern for carp include Saprolegnia spp. Achlya spp. and Branchiomyces spp., which cause necrosis of the gill filaments (gill rot).

Warm water fish hatcheries are particularly susceptible to water mold disease, and of importance for carp are members of the family Saprolegniaceae, especially the genus Saprolegnia. Dead or unfertilized fish eggs are very quickly infected by fungi, and left to proliferate, these quickly threaten healthy eggs and fry. The resulting economic losses in intensive aquaculture can be severe. Typical signs of a Saprolegnia infection (saprolegniasis) are cotton-wool-like white colonies of the fungus on the integument of the fish. Fungal hyphae penetrate the epidermis and the dermis, causing degenerative changes including edema and, eventually sloughing of the skin. Left untreated, infected fish die of osmoregulatory failure. Disinfection measures should be taken as soon the first signs of saprolegniasis are detected. It should be noted that the use of malachite green to treat fungal and other ectoparasitic diseases such as ichthyophthiriosis is no longer recommended, or in fact permitted, in fish farms within the European Union (EU). As an alternative, therapeutic baths with NaCl, formaldehyde or kaliumpermanganate are recommended. Other suggested therapeutics to cure saprolegniasis are very similar to those for columnaris disease.

Other fungal species commonly implicated in carp disease are Aspergillus, Penicillium and Fusarium, which can be ingested with contaminated food. Several pathogenic fungi are known to produce mycotoxins (e.g., aflatoxin from Penicillium sp.) that can be detrimental to a wide variety of animals including fish. However difficulties in detecting and identifying specific mycotoxins mean it is often impossible to prove a mycotoxicological cause for disease, and the picture is further clouded by the fact that fungi may be present without producing any toxins.

Diseases caused by Protozoa

The protozoans infecting fish include both ecto- and endoparasites. Among the endoparasitic species, blood flagellates of the genera Trypanosoma, Cryptobia and Trypanoplasma spp. are of special importance and the most troublesome with respect to carp is the trypanosome Trypanoplasma borelli. The ectoparasitic protozoans fall into three broad types. Firstly there are the opportunists with only limited persistence as parasites. These are typically secondary infectors that parasitize mainly stressed or immune-depressed fish. Examples include ciliates of the genera Ophryoglena, Tetrahymena, Hemiophrys and Glaucoma. Secondly there are the ubiquitous facultative parasites lacking host or site preference, such as the flagellate Ichthyobodo necator (formerly Costia necatrix), ciliates of the genus Chilodonella, Ichthyophtirius multifilis, Crypotcaryon irritans, several ubiquitous members of the Trichodina and Tripartiella species complexes such as Trichodina reticulata which infects the skin of carp, and parasitic dinoflagellates of the genus Oodinium (Piscinoodinium). Thirdly, there are numerous specialized facultative parasites exhibiting limited host specificity but with a pronounced preference for a specific infection site. These include the highly specialized trichodines and several Tripartiella spp., all of which parasitize the gills of fish.

One of the most common protozoan parasites is the ciliate Ichthyophtirius multifilis. Outbreaks can develop into devastating epidemics in intensive fish farms and ornamental fish facilities, often resulting in total stock losses. Commonly known as 'Ich' or 'white spot' the disease is typified by visible white spots on the body surface, and sometimes also on the gills. Each white spot represents an individual parasite in the trophont stage, during which the ciliate resides in the host's epidermis where it feeds on skin cells and body fluids and develops to maturity. It then leaves the host and falls to the bottom where it transforms into a round reproductive cyst, the tomont. The tomont undergoes repeated divisions until it has divided into 100 to 2000 ciliated, free-swimming infective stages known as theronts. The theronts search actively for a new host and on contact with a suitable fish (Paperna suggested that almost any freshwater teleost species will do) the theront penetrates the skin, burrows into the epidermis and transforms into the feeding trophont.

The time required to close the life cycle is temperature dependent, with generation time decreasing with rising water temperature. At 20 °C, the entire life cycle can be completed in just 8 days. In warm water aquaculture conditions or in closed systems such as aquaria, I. multifilis can cause 100% mortality within a very short time. Often the first sign of an I. multifilis infection is so called 'flashing' of fish. Infected individuals turn onto their sides and rub themselves on the substratum in an attempt to rid themselves of the parasites. Sunlight reflected off their upturned silvery flanks is seen as flashing in the water. However similar scratching behavior can also be a sign of other skin irritations, so care should be taken with diagnosis. The best approach is to search the skin and gills for the characteristic white spots and examine a skin smear under a microscope. In case of infection this will reveal individual ciliates (up to 1 mm in diameter) often rotating slowly due to the beating of their cilia. The nucleus of the trophont stage has a characteristic horseshoe-shaped macronucleus. Ichthyophthirios is one of the most troublesome parasitic diseases in cold- and warm water fish cultures of carp, eels and catfish, and among the most difficult to treat. Fish may sustain low enzootic infections and encysted tomonts may survive in the habitat for a long time. Stressed host fish kept at high densities during rising spring temperatures provide optimal conditions for the parasite, and increase the risk of epizootic outbreaks that result in vast losses in a very short period of time. However spontaneous recovery has been reported, both in natural fish populations and in fish holding facilities. Clearly some fish species are capable of mounting an effective defense against I. multifilis, including humoral and non-humoral immune reactions. In carp, limited immunity was achieved following a controlled exposure to I. multifilis tomonts, and fish subsequently survived high doses of the parasite. However this immunity was not stable and was suppressed following the application of corticosteroids.

Diseases caused by Myxozoa

The Myxozoa are spore-forming parasites that infect freshwater and marine fishes. Previously classified as protozoans, recent molecular analyses and observations of specific functional specializations and multicellular states have required their reclassification a separate phylum of metazoans including more than 2100 described species to date. Most members of the Myxozoa are not harmful to fish but a few individual species are highly fish-pathogenic. A common and economically important disease of carp in intensive aquaculture is swimbladder inflammation (SBI), caused by Sphaerospora dykovae. S. dykovae belongs to the class Myxosporea, a group of myozoan parasites that form distinctive infective spores easily identifiable under a microscope by their polar capsules containing coiled filaments. A typical myxosporean life cycle includes two obligate hosts, an annelid (oligochaetes in freshwater and polychaetes in marine water) and a vertebrate, typically a fish. In the latter, the typical myxosporean spores develop by sporogony and are released into the water. They are ingested by the annelid host, in which mature actinospores are formed via sporogony and released into the water. The actinospores infect the fish following contact with the skin, fin or gills and subsequent invasion of the sporoplast completes the life cycle.

The pathology of SBI in carp is typically divided into five stages. During the first stage, blood vessel dilatation leads to hyperaemia and a petechial rash in the swim bladder wall. In the second stage, hyperaemia decreases and brown or black spots can be seen in the swimbladder wall. In the third stage the swim bladder wall is thickened by inflammation and filled with an exudate. These processes are exacerbated in the fourth stage and layers of the swim bladder wall are necrotized. In the fifth stage cysts are formed in the swim bladder and the organ is filled with serum or pus. At this stage inflammation can spread to other organs and secondary peritonitis may develop. From the third stage onwards, secondary bacterial infections may exacerbate or add to the condition. During the first and second stages of SBI there are no outwards signs of disease and the pathology described above can only be detected by post mortem analysis. It is usually in the third to fifth stages that fish suffering from the disease become noticeably sick. In the chronic form, infected fish may lose their balance and swim on their back or side, or even tail-up. The abdomen is enlarged and the caudal fins may stand out of the water as the fish tries and fails to dive. However these clinical signs typically appear in only 10–20% of infected fish and mortality is relatively rare.

Most problems with SBI are in fact caused by accompanying secondary infections for example by bacteria. These can lead to severe losses in hatcheries where the disease can spread rapidly in carp fingerlings. According to Jeney and Jeney no treatment against sphaerosporosis is available as yet, and the only effective agent for prevention is fumagillin, which will prevent the development of the disease if added to fish feed.

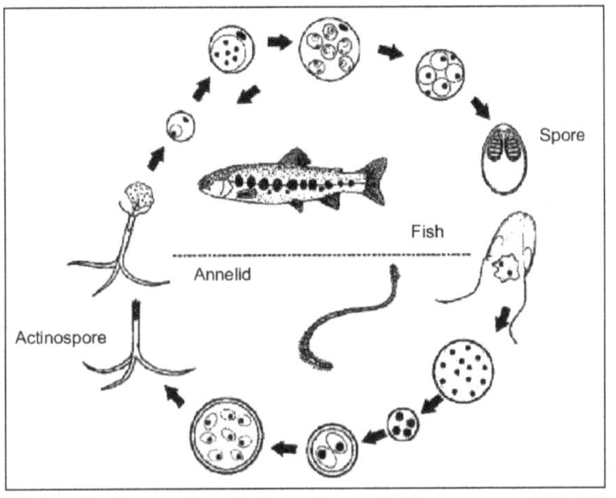

Typical myxosporean life cycle with alternating fish and annelid hosts, modified from that of Yokoyama et al.

Diseases caused by Platyhelminthes

The platyhelminthes are a species-rich group with three parasitic classes, the trematodes, monogeneans and cestodes.

Trematoda

Most trematodes have a complex life cycle including several obligate hosts. Depending on the location of infection in the intermediate vertebrate host, the various species are commonly referred to as eye, blood, liver or lung flukes. In most species, the final host is a vertebrate, such as a fish-eating bird like a gull or a cormorant. Within this final host, the trematodes generally inhabit the intestine, where they attain maturity and reproduce sexually. They produce eggs which are voided into the environment along with the feces of the host. Free-living motile miracidia hatch from the eggs and seek a first intermediate host, usually an aquatic snail, which they infect via penetration of the skin. Inside the first intermediate host, the parasite enters its next life stage, known as the redia or sporocyst, which reproduces asexually to produce large numbers of the next infective stage, the cercaria. Cercariae are shed continuously from the snail at a rate of several hundred to several thousand per day for up to several months. The cercariae of most species are motile, and search actively for the next life cycle host, which they usually infect by penetration of the skin. Depending on the species of fluke, this next life cycle host might be a second intermediate host within which the parasites develop into metacercariae that are infective for the next intermediate host, or it might be the final host within which the adult worm develops. The number of obligate intermediate hosts varies between one and six depending on species of fluke.

Eye flukes belonging to the Diplostomum species complex infect a range of fish species including carp and all its varieties. In the second intermediate fish host, metacercariae are found in the lenses or other parts of the eyes depending on the infecting Diplostomum species. In severe infections, the concentration of parasites in the lens can cause cataracts and the fish may go blind, as seen in rainbow trout Oncorhynchus mykiss infected with D. spathaceum. In fish fry, mass infection with cercariae of Diplostomum sp. leads to a disease known as diplostomiasis which can result in severe losses. Another trematode causing serious disease mainly in carp fry is the blood fluke Sanguinicola inermis. It infects a broad range of fish hosts, including all varieties of Cyprinus carpio and several other cyprinids, and other species such as perch, Perca fluviatilis, and pike, Esox lucius. The life cycle is relatively simple, including an aquatic snail intermediate host and a fish final host. The blood flukes cause gross histopathology in carp including mechanical damage caused by the movement of the flukes through the host's tissues and blood vessels and inflammatory reactions to encapsulated eggs. Inflammation and necrosis of the gill tissues may be a threat to cultured fish.

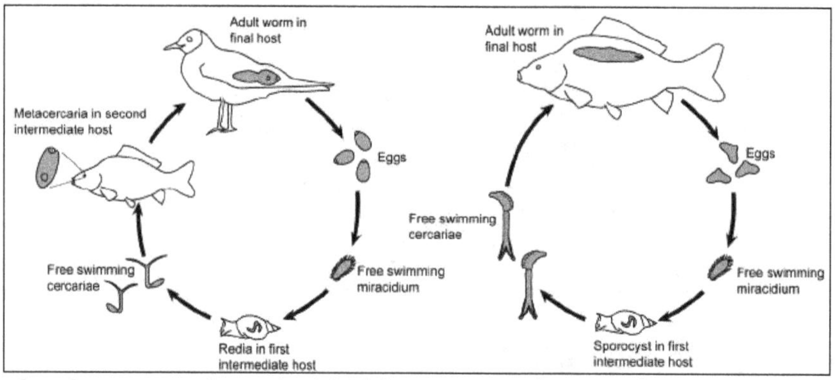

Life cycles of two trematode species (a) Diplostomum spathaceum, (b) Sanguinicola inermis.

Outbreaks of trematode disease in fish rearing ponds are best prevented by controlling the snail host population and by mechanical filtration to remove cercariae from circulation. The antihelmintic Praziquantel (also known as Droncit) can be used to cure infected fish.

Monogenea

The monogeneans constitute a very diverse group that infect freshwater and marine fish species almost exclusively. Owing to a long history of host-parasite co-evolution, monogeneans tend to be highly host-specific, and in the wild seldom cause any detriment to the fish they infect. However under the dense rearing conditions of aquaculture and the ornamental fish trade, members of the monogenean orders Dactylogyridea and Gyrodactylidea can be seriously problematic. Paperna recorded annual losses of carp fry to Dactylogyrus vastator infections in breeding and nursery carp ponds during the spawning season. Unlike the trematodes, most monogeneans have a simple, direct life cycle and spend their whole life on one single host. Members of the Dactylogyridea are oviparous and infect mainly the gills of their hosts, while the viviparous Gyrodactylidea infect the skin of their host fish. A range of effective treatments exist, including Praziquantel and formaldehyde.

Cestoda

The cestodes or tapeworms constitute a diverse group of approximately 3400 parasitic species with vertebrate hosts (including 800 known species infecting teleosts). Their large size (some species can grow to several meters of length) and conspicuous nature makes cestodes perhaps the best known and most reviled of parasites, and those infecting fish are generally perceived by farmers, anglers and fishermen to be detrimental. In reality however, most tapeworms dwell in the intestine of their hosts, attached by suckers, hooks or other holdfast organs and doing nothing more harmful than absorbing nutriment from the content of the hosts gut via their own skin. Other than a generally marginal reduction in food or vitamin uptake for the host, low numbers of cestodes do little harm and in most cases are probably imperceptible to the host. In commercial terms, cestodes are not usually a serious concern, though those that reside in the host's flesh can reduce demand and profitability. However there are exceptions. Heavy infections with Bothriocephalus acheilognathi and Khawia sinensis can result in mass mortalities of carp. Both species originate from Asia— B. acheilognathi from the Amur River and K. sinensis from China—and both have been introduced elsewhere along with commercial carp imports. Since arriving in Europe, B. acheilognathi has spread along with its carp host and now also infects more than 40 different cyprinids and some further species. In the USA, its presence is a threat to a number of endangered fish species including the woundfin minnow (Plagopterus argentissimus), roundtail chub (Gila robusta) and speckled dace (Rhinichthys osculus).

B. acheilognathi infects two host species during its life cycle. Eggs are released from adult worms that live in the intestine of the final host—a fish such as a carp. From these eggs, free-swimming oncomiracidia hatch and are consumed by the intermediate host, a copepod. The procercoid stage develops in the haemocoel of the intermediate host and infects the final host after it has ingested the copepod prey. Clinical symptoms include clumsy movements of fish, lingering at the water surface and emaciation. Parasites usually accumulate in the anterior part of the carp's intestine where they can cause a blockage, enlargement of the abdomen or in extreme cases rupture of the intestinal wall. The holdfast organ comprises two large sucking grooves, which may engulf parts of the intestinal wall, causing local inflammation and local loss of gut epithelium, sometimes followed by a drastic host immune response. Adult carp infected with tapeworms can be treated in spring and autumn with an antihelmintic (such as Niclosamid). After treatment, ponds should be emptied, dried and treated with unslaked or chlorinated lime.

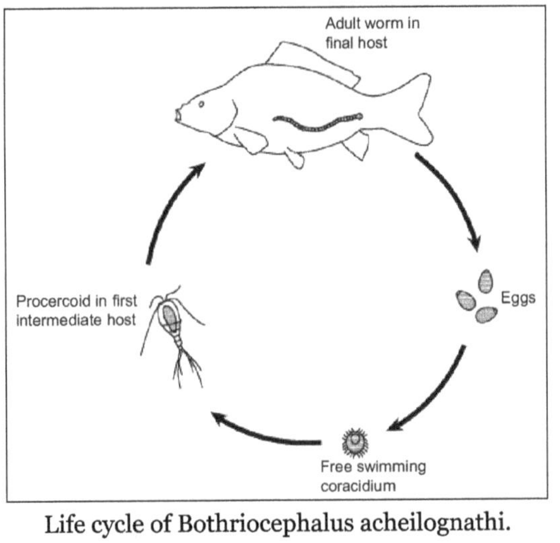

Life cycle of Bothriocephalus acheilognathi.

Diseases Caused by Nematodes

The vast phylum Nematoda includes a number of species that parasitize fish, all of them endoparasites. They occupy a variety of host tissues, but only a few species are known to be pathogenic to carp. These especially include members of the genus Philometroides such as Philometroides cyprini (formerly known as P. lusii or P. lusiana) in common carp and Philometroides sanguineus in crucian carp. Large infestations of either species may cause a disease known as philometroidosis in both cultured and wild fish. P. cyprini has a complex annual life cycle, in which the carp is the final host and various copepod species can serve as intermediate hosts. Gravid female nematodes occupy the skin under the scales of host fish and release infectious nematode L1 larvae in spring (May/June in Europe). If ingested by a copepod, the larvae go on to develop into the fish infective L3 larvae. When an infected copepod is ingested by a fish, the nematodes penetrate the gut wall and migrate into the body cavity. They aggregate near the swim bladder, gonads and kidney where they grow to 3.5 mm in length and copulate approximately one month post infection. The males then migrate to the swim bladder wall and the females to their position in the skin under the scales, where they continue to grow quickly and may reach up to 110 mm in length by spring. They produce numerous L1 larvae of approx. 450 μm in length.

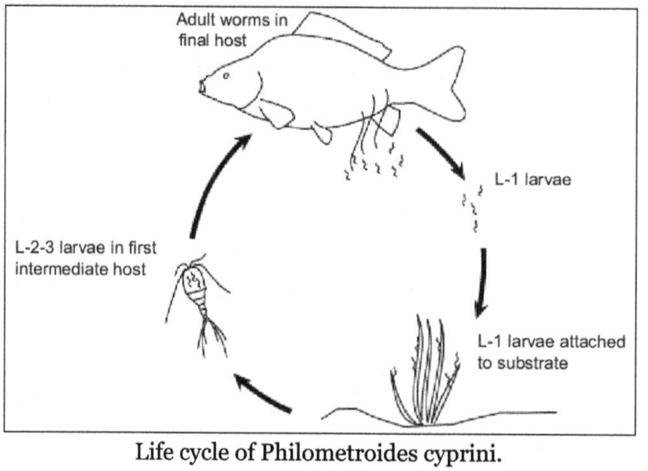

Life cycle of Philometroides cyprini.

P. cyprini parasites are pathogenic to the host. Migrating juveniles cause inflammation of the swim bladder wall and hematophagic feeding habits result in mechanical damage to blood vessels. The movement of female nematodes to the integument leads to the destruction of muscle fibers and increases the risk of secondary infection, for example with Saprolegnia spp. For juvenile carp (1^+), infection with more than three female nematodes may be fatal. Infected hosts exhibit reduced growth, lethargy and abduction of scales, that becomes apparent in spring. Infected fish should not be spawned in hatcheries. The best preventative action against philometroidosis in carp is stringent separation of year classes. Spawners should be removed from the breeding ponds immediately after spawning. Carp fry should be moved from the breeding ponds to nursery ponds no later than 6 days after hatching because the L1 nematode larvae are not infectious for carp for 7 days after release from the female.

Diseases caused by Crustaceans

The Crustacea (phylum Arthropoda) are an extremely diverse group comprising over 38,000 species. Most crustaceans are primarily aquatic and a large number have a parasitic association with fish. The majority of crustacean fish parasites are ectoparasites that feed on host blood, mucus or skin cells, many also parasitize fish gills. Pathology therefore tends to include mechanical disruption of the epidermis at attachment and feeding sites, often resulting in osmoregulatory or respiratory problems and creating opportunities for secondary infectors such as pathogenic fungi and bacteria. The most well-known crustacean parasite of carp is Argulus foliaceus, which despite its low host specificity is commonly known as the 'carp louse' due to its significant impact on the carp farming industry of Europe and Asia. The pathogenicity of the carp louse lies not only in mechanical damage to the fish's skin but also in irritations caused by the secretion of digestive toxins into the epidermis while feeding. Furthermore, it has been shown that A. foliaceus can act as a vector for other pathogens such as Rhabdovirus carpio, the agent of spring viremia of carp (SVC). A. foliaceus has a direct life cycle in which adult females leave the host after copulation and deposit their eggs (up to 500) in a gelatinous string on any suitable surface such as stones. The females may then return to the same host individual or attach to another until ready to deposit the next clutch of eggs. Incubation times for the eggs vary with temperature, but once hatched (within 8 days at 26 °C), the parasitic metanauplii search actively for a host fish. After attachment, nine larval molts follow in quick succession and the adult stage is reached. In warm summer temperatures, the whole life cycle may be completed in less than 40 days, making it possible for the parasite to realize four generations per year under optimal conditions. Reproductive success peaks at temperatures between 20–25 °C, but grinds to a halt at temperatures below 13 °C. Fish suffering argulosis begin to show behavioral abnormalities such as lethargy and loss of appetite shortly after the onset of infection. Heavy infections may lead to fin damage, increased mucus production and small petechial hemorrhages at parasite attachment sites that may result in skin loss, osmoregulatory distress and secondary infection by fungi and bacteria. The symptoms may lead to significant host mortality in fish farms, rearing ponds and natural habitats. Numerous chemotherapeutics are available to treat argulosis, most of them based on organophosphate insecticides. Ponds with infected stock should be dried and disinfected to eliminate the eggs of the parasite.

Asian Seabass

The Asian seabass (Lates calcarifer) is an important candidate finfish species for farming. Seabass is a euryhaline fish, growing rapidly up to 3-5 kg within a growing period of 2-3 years in both freshwater and brackishwater environments. For maturation and spawning it migrates to the sea while the postlarvae and juveniles migrate to lagoons and backwaters for growing. It is a voracious carnivorous fish. However, the juveniles are omnivorous, feeding mainly on crustaceans and other small fishes. Seabass attains maturity at the age of 3-4 years at a length and weight range of 60 to 70 cm and 2.5 to 4.0 kg respectively. Males are generally small and in the size range of 2.0-3.0 kg and the males convert into females as they reach a size above 5.0 kg. The fecundity is between 2.1 to 17.0 million depending upon the size of the fish.

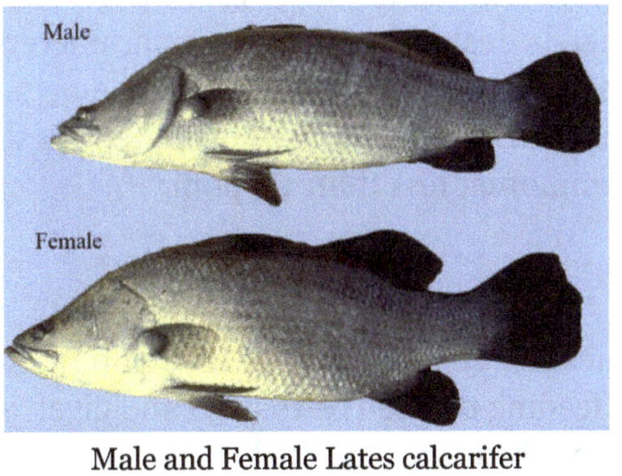

Male and Female Lates calcarifer

Seed Production

The required important facilities for a seabass hatchery complex are:

- Broodstock holding tanks,
- Maturation tanks,
- Spawning tanks,
- Egg incubation tanks,
- Larval rearing tanks,
- Artemia hatching tanks,
- Live feed (algae and rotifer) culture tanks, and
- Nursery rearing tanks.

The saline water can either be drawn from a borewell in the intertidal area or from open sea. Water drawn should be stored in reservoirs and filtered through biological filters, rapid sand filters and ultra-violet, filter to maintain the required water quality.

Broodstock Development and Maintenance

The source of the broodstock can be either from the wild or from rearing ponds/cages. Brood fishes range in size from 2.0 to 10.0 kg, smaller size for males and larger ones for females, which can be induced to mature and spawn within 6 months. Transportation can be done using open or closed containers lined with soft materials like rubber foam mattress. Before transferring the fishes into the broodstock holding tanks (12m x 6m x 2m or 7m x 4m x 2m; rectangular in shape: with PVC inlet and outlet), they should be acclimatized in the holding tanks for. 2-3 days.

Stocking

Stocking of 1 kg fish biomass/m3 is recommended for a 100 tonne of water, i.e. 10 females each of average weight 6.0 kg and 16 males each of 2.5 kg.

Water Quality Management

Water exchange to an extent of 70-80% of the total volume should be done daily. A flow through arrangement for water exchange is desirable. Important parameters of water quality to be maintained are as follows: water temperature: 28-32°C; salinity: 29-32 ppt; alkalinity (CO3): 80-120 ppm; pH: 6.8-8.0; dissolved oxygen: above 5 ppm; phosphate: less than 10 ppm; unionised ammonia: less than 5 ppm; ionised ammonia: less than 1.5 ppm;

Feed Management

Trash fishes like Tilapia/sardines can be procured, cleaned and packed as 2-3 kg blocks and frozen in deep freezers. At the time of feeding this can be thawed and given as feed to the broodstock @ 5% of the total biomass. Conditioning the broodstock fishes to feed on frozen trash fish should be done gradually, first with live and frozen fish and later with frozen fish only. Excessive feeding may be avoided and left-over feed should be removed immediately, since it will pollute the rearing medium.

Health Management

Parasites like Caligus sp., Lernanthropus sp. are common in seabass. The most problematic parasite in seabass broodstock maintenance is the monogenic trematode Diplectanum latesi. Infected fishes do not feed, become lethargic and swim in isolation. Bath treatments with 100 ppm formalin for crustacean parasites and 1 ppm organophosphorus pesticide Dichlorovos, for 1 hr, is effective in controlling the parasites.

Maturation

Seabass matures spontaneously in captivity during June to October. The maturation process can be accelerated by hormonal pellet implantation. The maturity stages of the broodstock should be monitored every fortnight. For assessing the maturity stage of female, ovarian biopsy is done by catheterisation. A polythene cannula of 1.5 mm diameter is inserted into the oviduct through the genital opening and the eggs are collected. When the diameter of eggs is more than 0.450 mm in size, the fish can be induced to spawn. In matured males, white creamy milt will ooze out when the abdomen is gently pressed.

Induced Spawning

Matured seabass can be induced to spawn by hormonal manipulation. The hormones like LHRH-a (Luteinizing Hormone Releasing Hormone analogue), HCG (Human Chorionic Gonadotropin), ovaprim, ovatide and carp pituitary extract can be used for induced spawning, of which, the LHRH-a hormone is known to give consistent results. Female and male fish in the ratio of 1:2 are selected for hormone treatment. Hormone is administered normally in the early hours of the day to facilitate the spawning in the evening hours of subsequent day. Full moon or new moon days are preferred for spawning. LHRH-a hormone is administered intramuscularly in a single dose @ 60-70 mg/kg body weight for females and 30-35 mg/kg body weight for males and the fishes are released into spawning tanks of 15-20 tonne capacity.

The abdominal swelling and the courtship behaviour of the fish externally indicate the ovulation response. Good water quality and aeration should be provided in the spawning tank. After 35-36 hrs of hormone injection, the fish spontaneously spawns. Seabass is a protracted spawner (i.e. ova are spawned in batches by the same fish). The same female can be induced to spawn 3 times in a season with an interval of 15 days. Fertilization is external and the fertilized eggs, which are transparent, measure 0.78-0.80 mm in diameter and float on the surface. The unfertilized opaque eggs sink to the bottom. The rate of fertilization may be 60-90%. From a single spawning, 0.75 to 1.5 million eggs can be obtained.

Incubation and Hatching

The floating fertilized eggs can be collected from the spawning tanks by a scoop net made of bolting cloth and stocked @ 80-100 nos./ litre in cylindro-conical shaped incubation tanks of 500 litre capacity. Hatching takes place 17-18 hrs after fertilization. Newly hatched larvae measure 1.4-1.6 mm in length and the healthy ones are transferred to larval rearing tanks.

Larval Rearing

The larvae and post larvae are reared at first in indoor tanks until they metamorphose into fry at about the 20th day after hatching. Larvae are stocked @ 30-40 nos./litre in fibre glass/RCC rectangular tanks of 4-5 tonne capacity. Larvae thrive on the yolk for the first 3 days. The 3-day old larvae should be fed with rotifers (Brachionus plicatilis) @ 2 nos./ml initially and gradually increased to 5 nos./ml till 8th day. From 9th day onwards, larvae should be fed with 5 nos./ml. rotifer and Artemia nauplii. Artemia nauplii density should be maintained @ 1 no./ml initially and increased to 3 nos./ml upto 15th day. From 16th day to 25th day, larvae should be exclusively fed with Artemia nauplii at a density of 4-5 nos./ml. All the feeds should be given in 4 doses at an interval of 6 hrs. After 20 days the larvae metamorphose into fry. The survival rate from hatching to fry stage is about 35-42%.

Nursery Rearing

Nursery rearing of seabass fry in ponds and cages to stockable juvenile size is essential before release into the grow-out ponds. Nursery ponds may range in size from 500-2000 m^2. A water depth of 50-80 cm is desirable. In a prepared nursery pond, fry of 1.0 to 2.5 cm size can be stocked @ 20-50 nos./m^2. Water exchange to the extent of 30% is required daily. Fry must be fed with supplementary feed of chopped and ground fish (4-6 mm size) @ 100% of the body weight, twice a day, in the first week. The feeding rate is gradually reduced to 60% and 40% during second and third week respectively. Though seabass prefers live fish food it could be weaned to trash fish within 5-7 days. The nursery-rearing period is about 30-45 days.

On 25th day, the fry measures 1.0 cm, which should be transferred to nursery tanks in the hatchery or nursery hapas at the farm site. Fry are stocked @ 1000 nos./m^3 in 4-5 tonne capacity tanks. The cooked and minced fish meat, made into small pieces of 1.5-2.5 mm, should be given as feed ad libitum during the nursery rearing. Grading (removal of shooter fish) should be done on alternative days to reduce cannibalism. In the case of rearing seabass in the hapas (2m x 2m x 1m size), fry should be stocked @ 500 nos./m^2. Feeding and grading should be done as in the case of tank rearing. The expected survival rate would be 80-86% with an average size of 1.25g in 30-35 days of rearing in the tank or hapa. Nursery cage size may range from 3 m (3 × 1 × 1 m) to 10 m (5 × 2 × 1 m) with a mesh size of 1 mm. Stocking density with fry of size range 1.0 - 2.5 cm may vary from 80-100 nos./m. Feeding schedule to be followed as in pond rearing. Cages should be checked and cleaned regularly. The fry on reaching a size of 5-10 g at the end of a rearing period of 30-45 days can be stocked in the grow-out system. Usually a survival rate ranging from 50-70% could be obtained.

Larval rearing tanks

Grow-out Systems

Traditional Culture

Extensive culture of seabass as a traditional activity is followed in the Indo-pacific region. In low lying coastal ponds, juveniles of assorted sizes collected from estuarine areas are introduced and fed with the forage fishes like tilapia, shrimps and prawns available in these ponds. These ponds receive water from adjoining brackishwater or freshwater canals or from monsoon flood. Harvesting is done after 6-8 months of culture. Since seabass exhibit differential growth, the size of the harvested fishes varies from 0.5 to 5.0 kg. Production up to 2 ton/ha/7-8 months has been obtained.

Cage Culture

The traditional culture of seabass can be improved by stocking uniform sized seed at specific density and feeding them with low cost trash fishes/formulated feed. Water quality is maintained through periodic exchange. Fishes are allowed to grow to marketable size and harvested. Seabass culture can be done in more organized manner as a small scale/large scale activity in both brackishwater and freshwater ponds and also in cages.

The size of the cage may be 50 m (5 x 5 x 2 m) with mesh size depending upon the size of the fish to be stocked (0.5 cm for 1-2 cm size fish, 1.0 cm for 5-10 cm size fish, 2.0 cm for 20-25 cm size fish and 4 cm for fish larger than 25 cm in size). In cage culture, both floating and stationary net cages are used.

Floating Cages

Floating cages can be set on coastal waters where tidal fluctuation is wide. The net cages are hung on GI pipe, wooden or bamboo frames. The cage is kept afloat by styrofoam drum, plastic carbuoy or bamboo. The most convenient dimension for a cage is that of a rectangle and a volume of 50 m^3 (5.5m × 6m × 2m). The cage unit is stabilized with concrete weights at each bottom corner. The cage unit has to be anchored to the bottom.

Floating cages for seabass culture

Stationary Cages

This type is fastened to wooden poles installed at its four corners. Stationary cages are usually set in shallow bays where the tidal fluctuation is narrow. The mesh size of nylon net would depend on the size of fish. Fingerlings should be transferred to nylon net (mesh size of 2.0 cm) for about 2 months of culture period and then they are moved to a cage net of 4.0 cm mesh size until harvest.

Stocking with seed of uniform size is desirable to avoid cannibalism. The stocking density may be about 40-50 fish/m initially. However after 2-3 months of rearing, depending upon the survival rate etc., it may be reduced to 10-20 fish/m. Periodic transfer of fish from one cage to another is essential in order to grade the fish and maintain uniformity in size. Feeding schedule is similar to that in pond culture. The cages should be cleaned regularly and periodically inspected for damage. A production level up to 15-17 kg/m can be obtained in cage culture over a period of 7-8 months.

Pond Culture

The two-week nursery reared fingerlings are suitable for pond culture. The production pond can have concrete walls and a soft bottom, ranging in area from 0.1 ha to a few ha, water depth of up to 2 m and salinity of 5-10 ppt is suitable. Seabass culture in ponds can be carried out either by poly-culture method or by feeding with low cost fishes like tilapia/oil sardines or with extruded floating pellets. The pond is at first dried, tilled, leveled and manured with raw cow dung @ 1000 kg/ha. If required, lime is added @ 50-200 kg/ha to maintain soil pH above 7. Urea @ 100kg/ha and super phosphate @ 50 kg/ha can also be added to enhance the algal bloom. Sea water/fresh water is then filled to a depth of 60-70cm in the pond. When the pond water becomes light green in colour indicating sufficient development of algae in the pond, forage fishes are introduced.

In pond culture, stocking with seed of uniform size (5-10 g), @ 3000-5000 nos./ha is desirable. Feeding of fish is carried out following two methods. In the first method, the fish are fed exclusively with chopped trash fish @ 10% of biomass twice daily (08.00 & 17.00 hrs) and reduced to 5% subsequently. In the other, the food is made available in the pond in the form of forage fish like Tilapia (Oreochromis mossambicus). Pelletized feed can also be given. In a well-prepared pond, manured/fertilized with raw cow dung @ 1000-1500 kg/ha and urea @ 100-150 kg/ha, Tilapia adults (male and female in the ratio 1:3) are introduced and reared for 1-2 months prior to stocking with seabass. To maintain natural food production for the forage fish, periodic manuring at fortnightly interval is done @ half the initial dose. 20% of pond water is exchanged on alternate days. Harvesting is done by draining the ponds or by using seine nets. Grow-out pond culture of seabass can yield a production of 2-3 tons/ha within a rearing period of 7-8 months.

Harvesting

For sea bass farmed in cages, harvesting is relatively straightforward, with the fish being concentrated into part of the cage (usually by lifting the net material) and removed using a dip net. Harvesting sea bass 'free-ranging' in ponds is more difficult, and requires seine-netting the pond or drain harvesting. After harvesting, the barramundi are placed in ice slurry to kill them humanely and preserve flesh quality. Fresh barramundi is generally transported packed in plastic bags inside styrofoam containers with ice. Fish are usually transported live in tanks by truck.

Disease Management in Seabass

Seabass is prone to diseases caused by parasites, bacteria, fungi and virus. But diseases and abnormalities due to environmental stresses and nutritional deficiencies have also been recognized.

Viral Disease

Lymphocystis Disease

Lymphocystis disease is commonly found in seabass raised in cages especially among juveniles 4–7 cm in total length. It has been observed at all temperatures in rather high salinity. A gross sign of the disease is massive enlargement of the cell within the dermis layer of the fish skin, which resembles the cauliflower disease. Transmission is from fish to fish.

Bacterial Diseases

Fish reared in intensive culture conditions are exposed to extreme environmental fluctuations, and they may be more sensitive to stress than wild populations.

Aeromonas Bacteria

Whenever fishes are exposed to environmental stress or injury, aeromonas causes serious outbreaks of homorrhagic disease with high mortality. Temperature, pH, high CO2 and DO depletion, decomposition products, and free ammonia in the water, all of these can be considered as possible factors for Aeromonas infection. When seabass are overcrowded and water salinity is low for long periods, the diseased fish caused by A. punctata could be observed. Gross signs and behaviours are usually shown by hemorrhage on the fin and tail. In a heavy case, erosion of tail and fin can be seen.

Vibrio Bacteria

Diseases caused by Vibrio sp. typically appear as ulcerative hemorrhagic septecaemia. The typical symptoms of vibrio disease include congeston of the fins, eccymoses and petechiae on the body surface and frequently, haemorrhages and ulceration of the skin and muscle tissue. The tissues surrounding the infected anus (the vent) are usually reddened and inflamed. Internally, there is congestion and haemorrhagia of the liver, spleen and kidney, frequently accompanied by the presence of necrotic lesions. The gut and particularly the rectum may be distended and filled with a clear viscuos fluid.The body is completely covered by a thick layer of mucous. Occasionally, small-unbroken lesions are present. There may be a reddening of the caudal funs and vent. Internal organs appear normal. Young fish die more rapidly than adults.The pathogenic vibrio which have been isolated from seabass include Vibrio parahaemolyticus, V. anguillarum and V. alginolyticus.

Columnaris Disease

Columnaris disease caused by Flexibacter columnaris is one of the diseases commonly found in juvenile seabass which are raised in water of low salinity during rainy and winter seasons. Gross signs are observed by saddle-shaped lesions in the mid-body position about the dorsal fin of the fish. The bilaterally symmetrical lesion appears as a fuzzy, pale yellow white plague, with dark margins, often eroding in the epidermis. Clinically, the condition may be chronic, acute or peracute. The gram negative, aerobic bacillus (about 12 um) can be isolated from the lesion of the diseased fish.

Parasitic Protozoa

Protozoans are probably the most important group of animal parasites affecting fish. Many reports from all over the world indicate great losses in fish culture caused by protozoans. Obligate parasites such as the ciliate ichthyophthirius and certain species of the cnidosporidians are responsible for many of these losses. Many species, which are considered as commensal protozoans, may become pathogenic under certain conditions. Environmental factors affect the susceptibility of fish to certain protozoans. Oxygen concentration and temperature are the factors affecting both hosts and parasites. Since many protozoans transfer from fish to fish through the water, fish population density is an important factor. Tremendous infestation of protozoans can occur in a relatively short time where fish populations are dense. Other factors, such as host size, age, host specificity, immunity, and the aforementioned influences of host condition also play an important part in the host reaction to invasion by protozoans. Most host reactions to invasion by protozoans are directed towards expelling or isolating the parasite.

Protozoans cause harm to fish mainly by mechanical damage, secretion of toxic substance, occlusion of the blood vessels, depriving the host of nutrition and rendering the host more susceptible to secondary infections. Some of the most common clinical signs are changes in swimming habits, such as loss of equilibrium, flushing or scraping, loss of appetite, abnormal colouration, tissue erosion, excess mucous production, haemorrhage, and swollen body or distended eyes.

Cryptocaryon Sp.

Cryptocaryon is a marine counterpart of the freshwater Ichthyophthirius species and similarly causes the white spot diseases in marine fish. Its morphology and life cycle is quite similar to that of the "Ich". The surface of invaded fish reveals white pustules or numerous minute, greyish vesicles which are nests of cilliates burrowing under the epidermis. They feed on the host's cells underneath the epithelium and cause heavy irritation resulting at first in excessive production of mucous and finally completely destroying the fine respiratory platelets of the gill filaments. On the skin, this parasitic protozoan causes considerable lesions resulting in destruction of large areas of the epidermis. Secondary infection may complicate the situation and the host dies. The incidence of Cryptocaryon sp. in seabass showed a distinct peak during low water temperature period, with a marked prevalence during February. This ciliated protozoan probably causes more damage to fish populations over the entire world than any other single parasite.

Trichodina Sp.

Members of the genus Trichodina with about 60 species described from marine fish and related peritrichous ciliates are the most common parasitic protozoans that are especially harmful to young fish. The species attach themselves to the gills of marine fish. More than half of juvenile seabass heavily infected with this parasite died. Trichodina also causes problems to crowded seabass in cages.

Clinical signs of trichodinosis include excess mucous production, flushing, debility and hyperplasia and necrosis of the epidermis. The fin may become badly frayed in heavily infected fish and this may be accompanied by sluggishness and loss of appetite. Excessive number on the gills of infected fish interferes with respiration.

Henneguya Sp.

Henneguya is a flaggelate found to attach mainly to the gills of seabass in cages. In heavy infections, it may be found in the skin. Gross signs are hyperplasia, bronchitis plus necrosis. Life cycle involves a free-swimming dinospore, which moves by means of 2 flagellae. It attaches to the host and transforms to a sac-like trophont, which has elaborate attachment mechanism. It feeds and grows and detaches from host, sinks to substrate where it encysts and produces dinospores.

Epistylia Sp.

Epistylis is another protozoan found in seabass especially in freshwater environment. Epistylis belongs to the sub-class Peritrichia, and is common ectocommensals; however, it occasionally turns pathogenic. It attaches to the fish with its stalks. This protozoan may be present at a variety of temperatures and its number may be large enough to cause a grey mat on the epithelial surface.

Parasitic Helminthes

Worm diseases with the possible exception of those produced by monogenetic trematodes have not yet appeared to be a serious problem in seabass culture. This is probably due in large part to their complex life cycle and the difficulty in completing such cycles in the culture system. Helminthis parasites, which have been found in seabass, include monogenetic trematode, digenetic trematide, nematide and acanthocephala.

Monogenetic Trematodes

Monogenetic trematodes can be observed throughout the year. Abundance of these parasites and their seasonal distribution have not been studied. It has been reported that temperature apparently plays an important role in determining outbreaks of certain parasitic Monogenera. Peak infections of monogenetic trematodes usually occur among young susceptible fish. Such behaviour is advatanegous for the spread of the organism in a fish population.

A monogenetic trematode which has been found to be a dominant species in the gills of seabass is Diplectanum latersi. Dactylogyrus sp. may be present but this has not been fully identified.

Digenetic Trematodes

Lecithochirium sp. was found in the intestine of seabass especially in wild fish. Incidence of infection was 86.0 percent and average parasite burden was 5.5 Another digenetic trematode which was commonly found in the intestine of wild seabass is Pseudometadena celebesensis. Its incidence of infection and parasite burden were 100 percent and 9.3, respectively.

Nematodes

Although many species of nematodes are found either as adults or larvae in fish, few have been implicated as serious pathogens of their hosts. In seabass, nematode of the genus Cucullanus was found more common in the gut of larger fish than in that of young fish.

Acanthocephala

Acanthocephalid worms, despite their fearsome-looking proboscis with its rows of hooks, have not been observed as serious pathogens of fish. The great majority of acanthocephalus in seabass are found as adults in the gut.

Parasitic Copepods

The parasitic copepods are among the most devastating of fish parasites. The mature female usually attaches to the fish and feeds on the host. After copulation the female matures and produces egg sacs while the male dies.

Caligus Sp.

Caligus sp. has caused big problems in cultured seabass. They attach to the gills, buccal and opercular cavities, occasionally on the skin and fins of the seabass. Heavy infections can cause mass mortalities especially in young fish.

Lernanthropus Sp.

Lernanthropus are found attached to the gill of seabass especially in cage cultured fish. Large numbers of this parasite can cause anaemia to the fish host.

Treatment

Treatment is usually in the form of chemotherapy, possible combined with some of the preventive measures listed above. Chemical control should be a "last resort" in disease control.

Chemical Prophylaxis

To treat the pond accurately, the volume of water in the pond must be known. To determine the volume of a pond, multiply the number of surface unit area of water by the average depth of the pond.

The following chemicals are often used in the treatment of various fish diseases:

- 15 ppm of formalin (which contains 37–40% formaldehyde). This should be new stock.
- ppm of malachite green (should be zinc-free) for Ich.
- ppm of potassium permanganate.
- 0.25 ppm of dylox (dipterex).

For small ponds, dilute the chemical in a bucket of water and distribute evenly over the pond using a dipper. In large ponds, the chemical should be mixed in a large drum and distributed evenly over the pond.

General Treatments

When treating fish, it is advisable to know the quality of the water because such things as pH and temperature greatly affect treatment results. Treat a few fish first and see how they react before treating the entire group. Use only the drugs and chemicals that have been cleared by an authorized agency for use on food fish. Here are some general treatments for specific groups of pathogens:

a. Viruses	No treatment known. Avoidance and prophylactic measures are best to prevent viral diseases.
b. Bacteria	Treated best by injection, or use of food additives or antibiotics
	Terramycin 2.5 – 3.0 g per 100 lb body weight (BW) per day for 10 to 12 days (A withdrawal period of 21 days is necessary before the fish are marketed).
	Sulfamerazine, Nitrofurans, Furazin, 10 g per 100 lb BW per day for 10 days (also Furozone and Furanace)
	Erythromycin, 4.5 g per 100 lb BW per day for two weeks.
	Potassium permanganate used as a wide-spectrum treatment at the rate of 2 – 3 ppm in ponds gives good results.
c. External Parasites	Formalin is the best treatment for protozoans and gill flukes. Effective rates are 15 – 25 ppm as a pond treatment and 100 – 250 ppm for one hour as a prolonged treatment. At temperature below 60 degrees F (15.6 °C) fish will tolerate a formalin concentration of 250 ppm for one hour. A rate of 100 ppm for one hour should be used at higher temperatures. Fish should be closely watched during the treatment period; if they show distress, put them back into freshwater. Oxygen depletion may occur a few days after treatment with formalin. Malachite green has been used successfully to treat cryptocaryon. A combination of 25 ppm formalin and 0.1 ppm malachite green can give excellent results against cryptocaryon. Malachite green has been used as a dip treatment at a concentration of 1 : 15,000 to control fungus on both fish and eggs Acetic acid at a 1 : 5 concentration has been used for 1 – 2 minutes to control external parasites. Dylox has been shown to be an effective control for anchor worm and crustaceans. It is also effective as a treatment for gill and body flukes but are not effective against protozoans. Concentration is 0.25 ppm. At high pH levels, treatment should be done with caution. Acriflavin has been used at 3–5 ppm to treat external parasites.
d. Internal Parasites	Most parasites inhabiting the alimentary canal can be controlled by the use of antihelminthis Di-N-Butyl tin oxide. The tin compound can be mixed in the food at the rate of 1 % and fed at 3 % of BW for three days.

Remember the following precautions during treatment:

- Know the accurate volume of the water.
- Know the percent active ingredient of the chemical.
- Evenly distribute the chemical.
- Use only the chemicals approved for use on aquatic animals for human consumption.

Trout

Elongate, fusiform body shape with 60-66 vertebrae, 3-4 dorsal spines, 10-12 dorsal soft rays, 3-4 anal spines, 8-12 anal soft rays, 19 caudal rays. Adipose fin present, usually with black edge. No nuptial tubercles but minor changes occur to the head, mouth and colour in spawning males. Coloration blue to olive green above a pink band along the lateral line and silver below. Back, sides, head and fins covered with small black spots. Colouration varies with habitat, size, and sexual condition. Tendency for stream residents and spawners tend to be darker with more intense colour, whereas lake residents are brighter and more silvery. Absence of hyoid teeth is the most easily distinguishing characteristic from cutthroat trout.

Rainbow trout is native to the Pacific drainages of North America, ranging from Alaska to Mexico. Since 1874 it has been introduced to waters on all continents except Antarctica, for recreational angling and aquaculture purposes. Production greatly expanding in the 1950s as pelleted feeds were developed. Trout fisheries are maintained, or culture practised, in the upland catchments of many tropical and sub-tropical countries of Asia, East Africa and South America. As a result, several local domesticated strains have developed (e.g. Shasta and Kamloops), while others have been arisen through mass selection and cross-breeding for improved cultural qualities.

Main Producer Countries

Many countries were reporting rainbow trout farming production. Some of them have relatively insignificant output in comparison to the production from the larger systems that are located in the primary producing areas in Europe, North America, Chile, Japan and Australia.

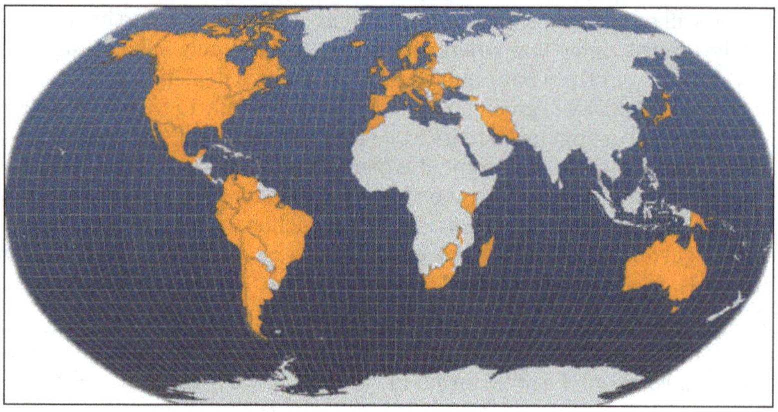

Main producer countries of Oncorhynchus mykiss.

Habitat and Biology

The rainbow trout is a hardy fish that is easy to spawn, fast growing, tolerant to a wide range of environments and handling, and the large fry can be easily weaned on to an artificial diet (usually feeding on zooplankton). Capable of occupying many different habitats, ranging from an anadromous life history [strain known as steelhead] (living in the ocean but spawning in gravel-bottomed, fast-flowing, well-oxygenated rivers and streams) to permanently inhabiting lakes. The anadromous strain is known for its rapid growth, achieving 7-10 kg within 3 years, whereas the freshwater strain can only attain 4.5 kg in the same time span. The species can withstand vast ranges of temperature variation (0-27 °C), but spawning and growth occurs in a narrower range (9-14 °C). The optimum water temperature for rainbow trout culture is below 21 °C. As a result, temperature and food availability influence growth and maturation, causing age at maturity to vary; though it is usually 3-4 years.

Females are able to produce up to 2 000 eggs/kg of body weight. Eggs are relatively large in diameter (3-7 mm). Most fish only spawn once, in spring (January-May), although selective breeding and photoperiod adjustment has developed hatchery strains that can mature earlier and spawn all year round. Superior characteristic selection is also achieved by cross breeding, increasing growth rates, resistance to disease, and prolificacy, and improving meat quality and taste. Genetic manipulation of the embryo sex chromosomes producing sterile, triploid females, hence avoiding the 'hook-like' jaw that does not appeal to the customer, and ensuring that introduced/escaped individuals cannot breed.

Trout will not spawn naturally in culture systems; thus juveniles must be obtained either by artificial spawning in a hatchery or by collecting eggs from wild stocks. Larvae are well developed at hatching. In the wild, adult trout feed on aquatic and terrestrial insects, molluscs, crustaceans, fish eggs, minnows, and other small fishes, but the most important food is freshwater shrimp, containing the carotenoid pigments responsible for the orange-pink colour in the flesh. In aquaculture, the inclusion of the synthetic pigments astaxanthin and canthaxanthin in aquafeeds causes this pink colouration to be produced.

Production

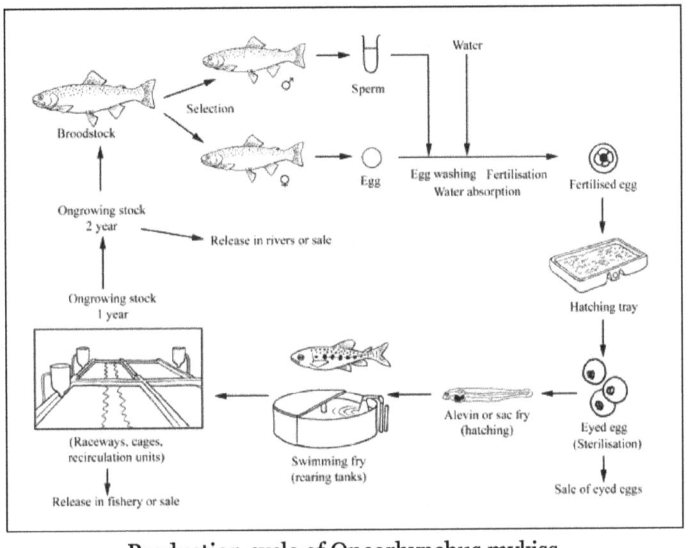

Production cycle of Oncorhynchus mykiss.

Production Systems

Monoculture is the most common practice in rainbow trout culture, and intensive systems are considered necessary in most situations to make the operation economically attractive.

A potential site for commercial trout production must have a year-round supply of high quality water (without aeration - 1 l/min/kg of trout without aeration or 5 l/sec/tonne of trout with aeration), that meets a number of criteria:

DO_2:	near saturation.
CO_2:	<2.0 ppm.
Temperature:	12-21 °C.
pH:	6.5-8.5.
Alkalinity (as $CaCO_3$):	10-400 mg/litre.
Manganese:	<0.01 mg/litre.
Iron:	<1.0 mg/litre.
Zinc:	<0.05 mg/litre.
Copper:	<0.006 mg/litre in soft water or <0.3 mg/litre in hard water.

Ground water can be used where pumping is not required but aeration may be necessary in some cases. Supersaturated well water with dissolved nitrogen can cause gas bubbles to form in the blood of fish, preventing circulation, a condition known as gas-bubble disease. Alternatively, river water can be used but temperature and flow fluctuations alter production capacity. Where these criteria are met, trout are generally on-grown in raceways or ponds supplied with flowing water, but some are produced in cages and recirculating systems.

Seed Supply

Development of Broodstock

Trout will not spawn naturally in aquaculture systems, hence eggs are artificially spawned from high quality brood fish when fully mature (ripe); although two-year-old trout start spawning, females are seldom used for propagation before they are three or four years old. The number of broodstock required is dependent on the number of fry or fingerlings required to meet the production schedule of the farm. The number can be back-calculated based on survival rates at the different life stages and the fecundity of the broodstock females. Generally, one male to three females is deemed a satisfactory sex ratio for broodstock. Males and females are generally kept separate. Broodstock maintenance can be costly and labour intensive, causing some farms to purchase eyed eggs from other sources; these should be 'certified disease free', although they should be treated with iodine (100 mg/litre for 10 min) upon arrival and gradually raised to the hatchery temperature. Broodstock are selected for fast growth and early maturation (usually after 2 years). One frequently used management tool is the use of sex-reversed, all-female broodstock to produce all-female progeny that grow faster. Functional males are produced by oral administration of the male hormone 17-methyl testosterone through starter feeds at the fry stage.

Stripping and Fertilization

The reproduction of rainbow trout is well understood and the techniques are well-developed. The dry method of fertilization without admixture of water is the most common approach. Eggs are removed manually from females (under anaesthetics) by applying pressure from the pelvic fins to the vent area or by air spawning, causing the fish less stress and producing cleaner, healthier eggs. Insertion of a hypodermic needle about 10 mm into the body cavity near the pelvic fins and air pressure (2 psi) expels the eggs. The air is removed from the body cavity by massaging the sides of the fish. Up to 2 000 eggs/kg body weight are collected in a dry pan and kept dry, improving fertilisation.

Males are stripped in the same way as females, collecting milt in a bowl, avoiding water and urine contamination. Milt from more than one male (ensures good fertilisation) is mixed with the eggs. It is recommended that milt from three or four males is mixed prior to fertilization to reduce inbreeding. Water is added to activate the sperm and cause the eggs to increase in size by about 20 percent by filling the perivitelline space between the shell and yoke; a process known as 'water-hardening'. Fertilised eggs can be transported after 20 minutes, and up to 48 hours after fertilization, but then not until the eyed stage (eyes are visible through the shell). Direct exposure to light should be avoided during all development stages, as it will kill embryos.

A technique that has been developed to improve production output is the use of monosex culture of females, or triploids. Triploidy is induced by exposing the eggs to pressure or heat whilst monosex are produced by fertilizing normal female eggs (XX chromosomes) with milt from sex-reversed, masculinised females (XXX chromosomes). The mature testes of sex-reversed fish are large and rounded but have no vent. The testes are removed from the abdomen and lacerated to drain the milt into containers. An equal volume of extension fluid is added to make the sperm motile, and ready for fertilizing normal ova. One advantage of this technique is that only the broodstock is sex-reversed, and they can be grown separately, while the marketed fish are not exposed to hormonal treatment.

Hatchery Production

Eggs are incubated undisturbed until the eyed stage is reached, in hatching troughs, vertical flow incubators or hatching jars. Hatching and rearing troughs are 40-50 cm wide, 20 cm deep, and up to about 4 m in length. They usually have 2 layers of eggs placed in wire baskets or screened trays (California trays) supported 5 cm above the bottom, and water passes through the tray (3-4 L/min). As the eggs hatch (4-14 weeks) the fry drop through the mesh to a bottom trough. The alternative is vertical flow incubators (Heath incubators) that stack up to 16 trays on top of each other. A single water source flows (3-4 L/min) up through the eggs, spills over into the tray below, thus becoming aerated, allowing large numbers of eggs to hatch in a minimal amount of space and water. Sac fry can remain in trays until swim-up at about 10 to 14 days after hatching. Time taken for hatching varies depending on water temperature, taking 100 days at 3.9 °C and 21 days at 14.4 °C (about 370 degree days). Hatching jars, available commercially or constructed from a 40 L drum and PVC pipe, introduce water from the bottom and flow from the top. 50 000 eggs can be incubated inexpensively suspended in a water flow that rolls the eggs, provided that the incubator contains two-thirds of the incubator volume in eggs, and the flow rate lifts the eggs 50 percent of their static depth. In all the above methods dead eggs are removed regularly to limit fungal infection. Fungal infections can be controlled using formalin (37 percent solution of formaldehyde) in the inflow water at 1:600 dilution for 15 minutes daily, but not within 24 hours of hatching. Upon reaching the eyed stage addling (dropping eggs 40 cm) removes weak and undeveloped eggs.

Trout hatch (typically 95 percent) with a reserve of food in a yolk sac (which lasts for 2-4 weeks), hence are referred to as yolk-sac fry, or alevins. Hatching of the batch of eggs usually takes 2-3 days, during which time all eggshells are regularly removed, as well as dead and deformed fry. Eggs incubated separately from rearing troughs are transferred to rearing troughs after hatching. After hatching, the trays are removed and trough water depth is kept shallow (8-10 cm) with a reduced flow until fry reach 'swim-up' stage, the yolk sac is absorbed, and active food searching begins.

Rearing Fry

Fry are traditionally reared in fibreglass or concrete tanks, preferably circular in shape, to maintain a regular current and uniform distribution of the fry, but square tanks are also found. Tanks are usually 2 m in diameter or 2 x 2 m square, with depths of 50-60 cm. Water is delivered to the side of the tank using an elbow pipe or a spray bar to create a circulation of water. The drain is in the centre of the tank and is protected by a mesh screen. This position ensures that the water forms a vortex towards the centre that accumulates wastes for easy removal. The sump or drain pipe is connected to an elbow pipe on the side of the tank that can be used to regulate water level.

Fry are fed specially prepared starter feeds using automatic feeders, starting from when approximately 50 percent have reached the swim-up stage. When most fish are actively feeding, 10 percent of the fish weight should be introduced daily for 2-3 weeks, preferably on a continuous basis using clockwork belt feeders. The feed pellets, made of fish meal (80 percent), fish oils and grains, provide nutritional balance, encouraging growth and product quality, and are formulated to contain approximately 50 percent protein, 12-15 percent fat, vitamins (A, D and E), minerals (calcium, phosphorus and sodium) and a pigment to achieve pink flesh (where desirable). High-energy commercial feeds and good feeding practices yield FCRs as low as 0.8:1. When the fry are 15-25 mm long feeding is based on published charts, related to temperature and fish size. Automatic feeders are useful but hand feeding is recommended in the early stages to ensure overfeeding does not occur, although demand feeders are more efficient for larger fish. As growth continues, dissolved oxygen is monitored and fish moved to larger tanks to reduce density.

Ongrowing Techniques

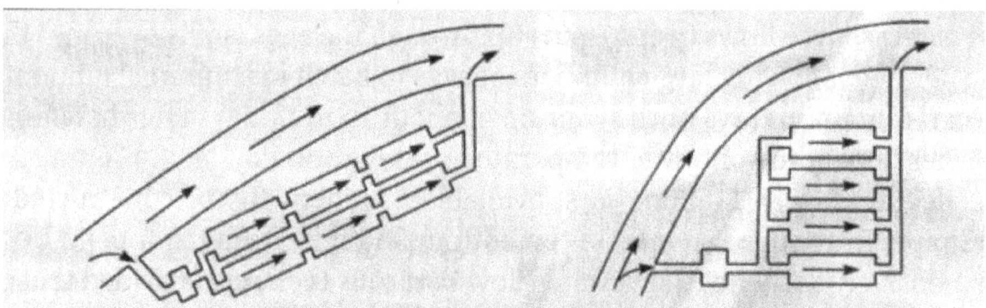

Raceway and pond layout.

When fry reach 8-10 cm in length (250 fish/kg) they are moved to outdoor grow-out facilities. These can comprise concrete raceways, flow-through Danish ponds, or cages. Individual raceways and ponds are typically 2-3 m wide, 12-30 m long and 1-1.2 m deep. Raceways provide well-oxygenated water and water quality can be improved by increasing flow rates; however, the stock is vulnerable to external water quality, and ambient water temperatures significantly influence growth rates. The number of raceways or ponds in a series varies with the pH [low pH (6.5-7.0) reduces unionised ammonia concentration] and the slope of the land (a 40 cm drop between each raceway is necessary for aeration). A typical raceway or pond layout is shown in above figure. For hygiene, water quality, and controlling disease problems the parallel design is better, as any contamination flows through only a small part of the system. Fry are stocked in both systems at 25-50 fry/m^2 to produce up to 30 kg/m^2 with proper feeding and water supply, although higher production is possible.

Fish are grown on to marketable size (30-40 cm), usually within 9 months, although some fish are grown on to larger sizes over 20 months. The stock is graded, usually four times (at 2-5 g, 10-20 g, 50-60 g and >100 g) in a production cycle (first year), when the density needs to be reduced, thus ensuring fast growth, improving feeding management and creating product uniformity. Fish quantity and size sampling (twice a month) allows estimations of growth rates, feed conversions, production costs, and closeness to carrying capacity to be calculated; essential considerations for proper farm management.

Alternative on-growing systems for trout include cage culture (6 m by 6 m by 4-5 m deep) production systems where fish (up to 100 000) are held in floating cages in freshwater and marine (past fingerling stage) environments, ensuring good water supply and sufficient dissolved oxygen. This method is technically simple, as it uses existing water bodies at a lower capital cost than flow-through systems; however, stocks are vulnerable to external water quality problems and fish eating predators (rats and birds), and growth rates depend on ambient temperature. High stocking densities can be achieved (30-40 kg/m^2) and fish transferred to marine cages have faster growth rates, reaching larger market size. Fry of about 70 g weight can attain 3 kg in less than 18 months.

Feed Supply

Feeds for rainbow trout have been modified over the years and cooking-extrusion processing of foods now provide compact nutritious pelleted diets for all life stages. Pellets made in this way absorb high amounts of added fish oil and permit the production of high-energy feeds, with over 16 percent fat. Dietary protein levels in feeds have increased from 35-45 percent and dietary fat levels now exceed 22 percent in high energy feeds. Feed formulations for rainbow trout use fish meal, fish oil, grains and other ingredients, but the amount of fish meal has reduced to less than 50 percent in recent years by using alternative protein sources such as soybean meal. These high energy diets, are efficiently converted by the rainbow trout, often at food conversion ratios of close to 1:1. Feeding methods vary for production systems. Hand feeding is suitable for small fish eating fine food. Mechanical feeders, driven by electricity or solar power, are frequently used to feed set amounts at set intervals depending on fish size, temperature and season. Demand feeders can be used for fish greater than 12 cm.

Harvesting Techniques

Methods of harvesting vary but water levels in the holding facilities are generally lowered and the fish netted out. In pens and cages, the fish are crowded using sweep nets and are either pumped from the holding pen alive and transported to the slaughter plant, generally by well boat, or slaughtered on the side of the pens. The whole process is carried out with the aim of keeping stress to a minimum, thus maximising flesh quality.

Handling and Processing

Fish intended for restocking for angling purposes are handled carefully and checked for fin quality, size and any external signs of disease before being put into a special pond to await transport. Fish destined for the table are killed humanely after similar, but less stringent, checks. Before slaughter, all fish should be starved for 3 days and, once killed humanely, the head should be left on; beheaded fish spoil more quickly. Rainbow trout are supplied to markets either fresh or frozen, and their shelf life is 10-14 days if kept on ice. Trout are marketed as gutted whole fish, fillets (often boneless), or as value-added products, such as smoked trout.

Production Costs

As with any business, rainbow trout farms aim to increase revenue and reduce expenditure. This can be done by using the best value feed/seed and materials, and achieving an efficient FCR. The average cost of production is between USD 1.20 and 2.00/kg. Running costs can start at USD 100 per 1,000 fry purchased at 6-8 cm and feed for one year from USD 1,000-1,400. Veterinary and medicine costs are from USD 50/tonne with transportation and sales commission about USD 500/tonne.

Production Statistics

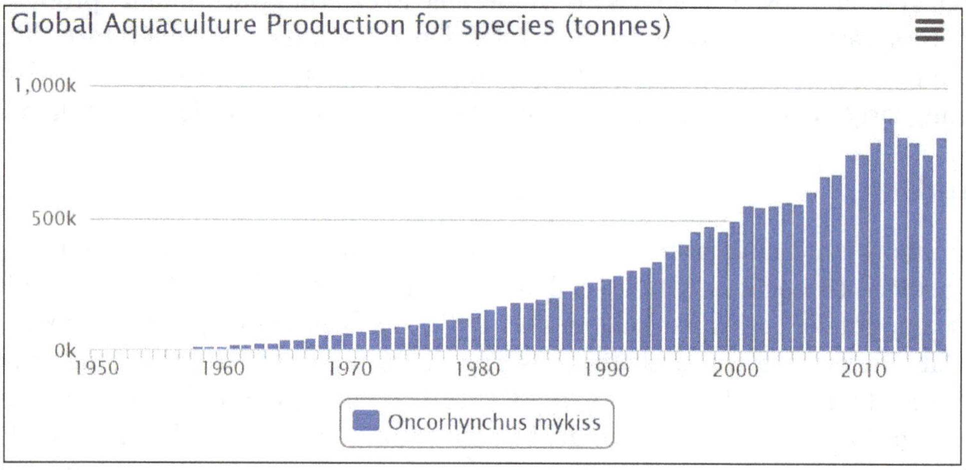

The production of rainbow trout has grown exponentially since the 1950s, especially in Europe and more recently in Chile. This is primarily due to increased inland production in countries such as France, Italy, Denmark, Germany and Spain to supply the domestic markets, and mariculture in cages in Norway and Chile for the export market. Chile is currently the largest producer. Other major producing countries include Norway, France, Italy, Spain, Denmark, USA, Germany, Iran and the UK.

Market and Trade

There are many outputs from rainbow trout culture, which include food products sold in supermarkets and other retail outlets, live fish for the restocking of rivers and lakes for recreational put-and-take game fisheries (especially in the USA, Europe and Japan), and products from hatcheries whose eggs and juveniles are sold to other farms.

Products for human consumption come as fresh, smoked, whole, filleted, canned, and frozen trout that are eaten steamed, fried, broiled, boiled, or micro-waved and baked. Trout processing wastes can be used for fish meal production or as fertiliser. The fresh fish market is large because the flesh is soft and delicate, white to pink in colour with a mild flavour. Food market fish size can be reached in 9 months but 'pan-sized' fish, generally 280-400 g, are harvested after 12-18 months. However, optimal harvest size varies globally: in the USA trout are harvested at 450-600 g; in Europe at 1-2 kg; in Canada, Chile, Norway, Sweden and Finland at 3-5 kg (from marine cages). Preferences in meat colour also vary globally with USA preferring white meat, but Europe and other parts of the world preferring pink meat generated from pigment supplements in aquafeed.

Strict guidelines are in place for the regulation of rainbow trout for consumption with respect to food safety. Hygiene and safe transportation of fresh fish are of paramount importance, to ensure that fish are uncontaminated by bacteria, in accordance with food agency directives.

Status and Trends

The rainbow trout farming industry has been developing for several hundred years, and many aspects are highly efficient, using well-established systems. However, current research and development is continually attempting to increase production efficiency and sales by increasing rearing densities, improving recirculation technology, developing genetically superior strains of fish for improved growth, controlling maturation and gender, improving diets, reducing phosphorous concentrations of effluents, and developing better marketing. One method that has been developed is a genetically modified hormone that is effective in reducing production costs. However, problems may lie ahead as public opinion towards genetically-modified products continues to be negative. As production continues to rise research is needed to keep costs to a minimum so the industry can move forward.

Main Issues

Trout farms inevitably impact upon the environment as river water is diverted from its natural course, potentially altering species composition and diversity. Escapee trout from farms can have negative impacts, potentially displacing endemic species (especially brown trout), and exhibiting aggressive behaviour that results in the altering of fish community structure.

Impacts from flow-through systems are largely from disease treatment chemicals, uneaten feed and fish excreta, which can alter water and sediment chemistry downstream of the farm. Elevated nutrients reduce water quality (increasing biological oxygen demand, reducing dissolved oxygen and increasing turbidity) and increase the growth of algae and aquatic plants. Output restrictions require farms to have settling areas to remove solid wastes, though soluble phosphorous in the effluent cannot be removed economically - hence reductions in feed are needed to address the problem. There are also problems with the transmission of diseases from farmed stock to vulnerable wild populations.

Diseases of Trout

Bacterial diseases that have been reported to cause recurring problems in recirculating systems are:

Bacterial Gill Disease

Bacterial gill disease (BGD) is one of the most common diseases of cultured salmonids. Mortalities can be high if fish are not treated early in an epizootic. The causative bacterium is Flavobacterium branchiophilum, and the disease has been experimentally transmitted with this bacterium. Although stressors are not required for disease outbreaks, they probably contribute to the number and severity of BGD outbreaks. Fish appear lethargic and may stop eating, they swim near the surface, and have distended opercles. Gills appear inflamed or with whitened tips, and excess mucus may be present. At the Freshwater Institute recirculating system, which has a continuous production schedule, BGD has been a recurring problem.

Table: The unit processes typically used in recirculating aquaculture systems, their action on water quality, and the resulting effect on fish health.

Component: Purpose	Significant conditions within the unit and their action on water quality	Importance to fish health
Culture tank: To contain fish during grow-out, allow fish to feed, and flush feces	• Fish respiration reduces levels of dissolved oxygen and increases levels of ammonia and carbon dioxide, which reduces pH. • Acid-base equilibrium (i.e., pH) controls the fraction of unionized ammonia and carbon dioxide present. • Feed introduces solids into the system. • Excretion produces particulate and dissolved solids. • High fish densities. • Fish may or may not carry pathogens. • Tank cleaning and fish handling or grading may occur that involve physical interaction with the fish.	• High fish densities and the deterioration in water quality (low dissolved oxygen and elevated carbon dioxide, ammonia, nitrite, and solids levels) may stress fish. • Physical interactions with the fish may produce fish stress or physical damage. • Horizontal pathogen transmission. • Elevated levels of ammonia and solids may induce gill pathology.
Clarification: To remove settleable and suspended solids via settling, sieving, flotation, or filtration.	• Most conventional clarifiers do not remove colloidal solids (~20 pm) which can produce anoxic or anaerobic conditions, exert an oxygen demand, support microbes, and leach nutrients.	• Fish pathogens may be associated with solids. • Some clarifiers store organic solids.
Biofiltration: To provide surface area where micro-organisms can establish; when the reused flow passes across these surfaces, the microbes remove a portion of the dissolved wastes.	• Microorganism metabolism lowers oxygen levels and produces carbon dioxide, which reduces pH. • Oxygen must be present for bacteria to oxidize ammonia and nitrite. • Increased levels in nitrite leaving the biofilter can occur if the biofilter is overloaded. • Biofilters may take 3-6 weeks to develop the population of bacteria to convert ammonia to nitrite and an additional 2-4 weeks to develop the population of bacteria required to convert the nitrite to nitrate. • Microbial growth produces solids within the biofilter. • Biofilm and biosolids produced in the biofilter may slough and be carried out with the recirculating flow, which contributes to the microbial counts and the colloidal concentration in the water.	• Biofilms may support certain fish pathogens, which may be passed to fish in the culture tank with sloughed biofilm canied in the reused flow. • Certain anaerobic by-products (e.g., sulfides) are toxic to fish.

Stripping/aeration: To contact water with air at near atmospheric pressures.	• Shifts concentrations of dissolved carbon dioxide, nitrogen, and oxygen towards equilibrium (i.e., adds oxygen, removes carbon dioxide and gas supersaturations). • Dissolved ozone can be stripped. • Strips link ammonia at the recommended pH levels (pH < 9.0).	• Elevated carbon dioxide reduces the capacity of blood to transpott oxygen and may induce nephrocakinosis.
Oxygenation: To contact water with purified oxygen at pressures generally 2 atmospheric.	• Generally used to create super-saturations of oxygen. • High total gas pmssures can be produced in units that do not vent off-gases. • Little carbon dioxide is removed due to insufficient gas exchange with respect to the volume of water treated.	• Gas supcrsaturations can produce gas bubble disease. • Increased oxygen levels can support much higher fish loadings in the culture tank, which adds stress.
Ozonation: To oxidize constituents in the water.	• Oxidation can reduce levels of nitrite, organic matter, microbes, water color, odor, or off-flavor compounds. • Oxidation can cause dissolved organic matter to precipitate and suspended organic matter to microflocculate, which enhances their removal. • Recirculating systems contain levels of organic matter and nitrite that teact with ozone and make sustaining an ozone residual difficult. • Ozone and certain of its by-products are toxic. • Accumulation of ozone and its oxidizing by-products is dynamic and ozone concentrations can increase suddenly and unexpectedly.	• Improved water quality reduces fish stress, which makes fish more resistance to disease. • Disinfection reduces risk from infectious diseases. • Ozone is toxic at low levels.

Table: Suggested water quality criteria for optimum fish health in freshwater.

Water quality parameter	Piper et al. on trout and salmonids	Laird and Needham on Atlantic salmon	Losordo on general guidelines for recirculating systems
Max. concentration carbon dioxide un-ionized ammonia nitrite nitrate total suspended and settleable solids ozone	10 mg/L 0.0125 mg/L as N 0.03 mg/L as N 3.0 mg/L as N 80 mg/L 0.005 mg/L	10 mg/L 0.0125 mg/L as N 0.02 mg/L as N 80 mg/L 0.005 mg/L	20 mg/L 0.02-0.5 mg/L as N 0.2-5.0 mg/L as N 1OOOmg/L as N
Optimum range PH dissolved oxygen	6.5-8.0 5.0 mg/L to satur.	5.5-8.5 >6 mg/L	6.0-9.0 >6.0

Infectious diseases that have been encountered in rainbow trout culture in recirculating systems:

Diseases

- Bacterial:
    - Bacterial gill disease
    - Furunculosis
    - Bacterial kidney disease
    - Fin rot.
- Parasites:
    - Gyrodactylus
    - Chilodonella
    - Trichodina, Epistylis, and Ttichophrya lchthyopthirius
    - lchtyobodo
    - Proliferative kidney disease
    - Amoebic gill infestation
    - Coleps
- Fungus:
    - Saprolegnia
- Viral:
    - Infectious pancreatic necrosis
    - Viral hemorrhagic septicemia
    - Infectious hematopoietic necrosis.

Each time new fingerlings were added to the system a BGD outbreak occurred within 6-8 days. Fingerlings were treated with Chloramine-T and mortality subsided until new fingerlings were stocked. E branchiophilum was not detected during quarantine, although some hyperplasia of the gills was present. Histologically, gills showed focal hyperplasia, areas of lamellar fusion and telangiectases. One of the factors which may have contributed to the outbreaks was the design of the cross-flow culture tanks which allowed for a continuous presence of suspended solids from the biofilter, feces, and uneaten food.

Only rainbow trout newly added to the reuse system became infected while fish residing in the system which had recovered from BGD, did not experience additional outbreaks. This suggests that some form of acquired immunity developed in fish which had survived a BGD outbreak and that they may have acted as carriers, transmitting the bacterium to the newly stocked fish. BGD must be diagnosed and treated promptly in a recirculating system in order to control the disease before massive mortalities occur. Mortalities can be controlled with Chloramine-T at 9-12 ppm for 1 hr or with benzalkonium chloride at 2 ppm repeated three times every other day for three treatments.

Other recirculating systems have had outbreaks of BGD, but only on a sporadic basis. Circular tanks were used in these systems rather than cross-flow tanks which reduces the suspended solids in the tanks. Also, the use of batch culture, where all fish are harvested from a tank before new fish are introduced, may have prevented carry-over of BGD in these systems.

At the Glenwood State Fish Hatchery, Utah, rainbow trout were raised from 10-26 cm in a recirculating system with rectangular circulation rearing ponds. Bacterial gill disease was a problem. The Chloramine-T and methylene blue (1 ppm) treatments used did not seem to affect biofilter function. Benzalkonium chloride (2 ppm) treatments had a negative impact on the biofilter. In this system, there were problems with biofilter function related to the cold water temperature which reduced Nitrobucrer reproduction; consequently, high nitrites were common. The rainbow trout in this system never seemed to flourish. Gill pathologies not associated with bacteria were also common.

Dworshak National Fish Hatchery (NFH), Idaho, has two recirculating systems for culture of B-run (a sub-species unique to the Clearwater river) summer steelhead trout (o. mykiss) from fingerling to smolt. The hatchery is supplied with surface water from the North fork of the Clearwater River, which contains several species of fish. The systems are comprised of rectangular circulation rearing ponds with submerged upwelling biological filter beds. The biofilter bed in one system uses polyethylene bead media (3/8 in long x 1/4 in diameter), while the other has Kochm biorings (3.5 in diameter). Reuse water from both systems is combined with 10% fresh water and is passed through an aeration tower before returning to the ponds. The recirculating systems are operated during winter months (late November through early-April) when river water temperatures are usually < 4 °C. The recirculating systems were designed so that water could be heated to 12-13 °C allowing for higher growth rates than could be achieved with ambient river water. BGD has not been a major problem, but outbreaks have occurred when the recirculating systems were overloaded (DI > 0.5 Piper et al.,) at the end of the grow-out period, resulting in high levels of ammonia. Gill pathologies not associated with bacteria were more common, and appeared when water quality characteristics exceeded the "no effects limits" for short periods or when there was chronic exposure to moderate levels of ammonia.

Schlotfeldt reported different degrees of gill pathology in rainbow trout cultured in recirculating systems in Northern Germany. The gill damage ranged from primary gill swelling (often due to chemical irritation with subsequent secondary bacterial infections), to different degrees of hyperplasia and finally to gill necrosis. The majority of gill necrosis cases were dependent on environmental factors, including high loads of organic ammonia and other catabolites which lead to ammonia autointoxication mechanisms. The lesions were reversible in most of the cases in which environmental conditions could be significantly improved, if they had not reached a critical degree of damage.

Furunculosis

Furunculosis is a bacterial infection of salmonids and other fishes caused by Aeromonas salmonicidu. The disease is characterized by a generalized bacteremia with focal lesions in the muscles.

Peracute infections occur in fingerlings, which turn dark and die with no other clinical signs except for slight exophthalmos. In acute infections, fish darken, stop feeding, and hemorrhage at the base of the fins. Internal hemorrhages occur and there are pathological changes in internal organs. Chronic cases usually occur in older fish with the presence of furuncles under the skin. Internally, they may have pus-filled lesions in the kidney, the lower intestine is often inflamed and there is a bloody discharge from the vent.

An outbreak of furunculosis at the Freshwater Institute in March, 1994 caused mortalities of several fish per day and was probably stress related. The ozone being added to the system had been abruptly turned off and resulting nitrite levels were high (0.46 mg/t.) for several days prior to the outbreak. A few fish started dying each day with clinical symptoms of furunculosis, including furuncles and swollen kidneys. Diagnosis was confirmed by isolation of A. sulmonicida from the kidneys. Two 5 day sulfadimethoxine/otmetoprim (Rome™) treatments at 50 mg/kg fish/day were necessary to control mortalities.

Subsequent samplings from the system have demonstrated the presence of the bacteria in the mucus but not in the kidneys of asymptomatic fish. The bacterium was not isolated from the biofilter or tank water after the outbreak ended, possibly because the bacterium did not become established in the biofilter or recirculating water of the culture system or it did not survive for extended periods of time in the system. The presence of Rome™ in the system may have reduced the numbers of A. salmonicidu. Failure to detect the organism could also be partially due to either the bacterium being overwhelmed by other hererotrophic bacteria or inadequacy of the detection procedures. A. salmonicidu has been isolated from the kidneys of sick fish which are seen occasionally in the system to the present day. Because of the continuous production method utilized, bacteria have become established in carrier fish. Another outbreak of funmculosis which required treatment occurred in May, 1996, with a few fish showing clinical signs each day. Our experience indicates that once this bacterium becomes established in the fish, clinical cases can recur at any time, usually related to a stressful event.

Fingerlings were quarantined upon arrival at the Freshwater Institute and checked for parasites and bacteria. A. salmonicida was never isolated from the kidneys of fingerlings in quarantine during five years of sampling. It appears that carrier fish were not detected by this method, allowing the organism to enter the system with the fish. No clinical disease was observed until the fish were severely stressed. Observations of other investigators have indicated that normal bacteriological examination of kidney samples is often inadequate to ensure reliable detection of A. salmonicidu among asymptomatic carrier fish. This situation highlights the importance of prevention and the need to improve detection methods for asymptomatic carriers. Munro and Fijan reported that furunculosis in rainbow trout in European recirculating systems was a recurring problem that was often serious. Information on mortalities and treatments for these epizootics was not reported.

Bacterial Kidney Disease

Bacterial kidney disease (BKD), caused by Renibacterium salmoninarum, is a chronic, progressive disease with potentially high rates of mortalities that is difficult to control and almost impossible to cure. The pathogen can be transmitted horizontally from fish to fish and vertically from broodstock to their progeny. Infected fish develop localized lesions (cysts filled with bacteria, pus and debris) in the kidney, liver, spleen, and heart. Once the disease reaches this state, treatments have little effect. Prevention of vertical transmission is an important tool in reducing the incidence and severity of BKD.

The Lahotan NFH raises Lahotan cutthroat trout (o. clarki) in a recirculating system with 60-70s water reuse. The fish grow in the system from 10 to 15-20 cm. During routine fish health monitoring, a very low level of BKD has been detected by the ELISA method, but no clinical disease has been observed. The broodstock are known to be carriers of BKD.

Munro and Fijan described BKD as a disease whose presence was continuous, and it was an intermittent problem in European recirculating systems producing rainbow trout.

Fin Rot

Fin rot is associated with poor sanitary culture conditions. Overcrowding and holding in raceways may lead to fin fraying with secondary bacterial infection. Bacteria such as A. salmonicidu and A. hydrophila, as well as flexibacteria, have been found in lesions of fin rot. In typical fin rot, fins become opaque at the margins, progressing towards the base. Epithelial hyperplasia follows, and in advanced cases the fins are frayed. Anderson reported fin rot in rainbow trout cultured in a recirculating system using a trickling filter. The system was treated with formalin at 1:4000 for 1 hr, which caused a temporary increase in the flushing of solids from the filter, but did not affect biofilter function.

Rainbow trout cultured at Glenwood State Hatchery, Utah, had fin rot infections with facultative aeromonads, such as A. hydrophila, which were probably caused by the stress of high nitrites present and the unhealthy condition of the fish reared in this system. Fin nipping due to aggressive behavior of steelhead trout at Dworshak NFH caused lesions in the dorsal fins where opportunistic microorganisms such as Aeromonus sp. and Pseudomonas sp. have been isolated.

Infectious Diseases-parasites and Fungi

Several parasites have caused problems at the Freshwater Institute, and other facilities as well, and a fungal infection was reported in Canada. It has been generally assumed that the parasites entered the systems with infected fish.

Gyrodactylus

Gyroductylus appears to be one of the most common parasitic problems in rainbow trout in recirculating systems, causing mortalities if a heavy infestation occurs. Gyrodactylus infests the skin and fins, multiplying when water quality conditions are less than optimal. Heavily infested fish may appear listless, have frayed fins, and have flashing behavior. The problem is easily controlled with a formalin treatment at 167 mg/L for 1 hr. Infestations with Gyrodactylus were present on several lots of fish at the Freshwater Institute during the quarantine period. Fish were treated before entering the recirculating system.

Gyroducrylus became a problem in the recirculating system on one occasion, and was successfully controlled with formalin. At Dworshak NFH Gyroductylus, has been isolated from steelhead trout during the recirculating phase, but has not been a major problem and treatment has not usually been necessary. When treatment was required, 150-200 ppm formalin for 4540 min has been effective.

Chilodonella

Chilodonella is a small, oval, colorless ciliated protozoan that infests the skin, fins, and gills of salmonids and warm water species. A mixed infection of Chiloubnefla and Ichryobodo was reported in a recirculating system. Three 24 hr treatments with 25 ppm formalin at two day intervals, eliminated Ichryobodo but not Chilodoneflu. A subsequent treatment with 200 ppm formalin for 1 hr was successful in controlling Chilodonella.

Trichodina, Epistylis and Trichophyra

Trichodina is a ciliated protozoan that can infest the gills, skin, and fins. This parasite was found at the Freshwater Institute system in conjunction with Gyrodacrylus or alone. Treatment with 167 mg/L formalin for 1 hr has been successful in controlling the parasite. Infestations occurred when water quality deteriorated. Trichodina have been counted in numbers as high as ~1500 liter in the system water at the Freshwater Institute without causing disease.

Epistyfis, Trichophyru, and Trichodinu have been observed from steelhead trout at Dworshak NFH and treated when necessary with formalin.

Ichthyopthirius

Ichthyopthirius mulrifiliis (Ich) is a large ciliated protozoan that infests the skin, fins, and gills. Infested fish have small gray-white pustules on the skin and fins, and will exhibit abnormal behavior, such as scraping their body against the bottom or sides of the tanks. Treatment of this parasite is difficult if the life cycle of the parasite becomes established in the recirculating system. Only the stages of the parasite that are free in the water can be successfully treated. The theronts, before they enter fish and the adult parasite leaving the fish are killed by formalin treatments, while the encysted stages cannot be killed. Repeated treatments with formalin at 167 mg/L for 1 hr are usually necessary.

Ich has been a problem at some recirculating facilities. In one system, a formalin treatment with 25 ppm for 24 hr, and two treatments with 35 ppm were not effective. Four additional 1 hr treatments with 167 ppm formalin every three days did not control the infestation. Subsequently, copper sulfate at a concentration of 0.15 ppm for 24 hr was used. Water alkalinity was 20 mg/L (as calcium carbonate). This treatment was repeated every third day, three times, and did control the parasite. In another system, Ich was controlled by an initial 24 hour treatment with 0.75 ppm copper sulfate, followed by a continuous treatment with 0.5 ppm for 30 days. Water alkalinity was increased to 75-120 ppm with sodium bicarbonate. Ich was not observed after 20 days, but treatment was continued for an additional 10 days. The fish were fed during treatment, and no detrimental effect to biofilter function was detected. At another facility, several days of chlorine disinfection of the system were necessary to prevent reinfection of newly introduced rainbow trout.

A mixed infestation of Ich and Trichodina in a reuse system with rainbow trout was treated with formalin but the treatments were ineffective. The fish were removed, the system was flushed, and restocked with fish from the same lot. The parasites multiplied rapidly again, and treatment was ineffective. The fish were removed again, and the system flushed without chemical treatment. The system was restocked with disease-free fish and the parasites did not reappear. The water temperature of 15.6 °C in the reuse system probably caused rapid multiplication of the parasites and high fish mortality.

The main infectious disease encountered at Dworshak NFH during the recirculating phase of rearing has been Ich. The parasite has been controlled by administering prolonged low-level formalin treatments (generally 35 ppm for 24 hr) to the entire system. This treatment regimen affected biofilter deamination-nitrification efficiency, requiring 2-7 days for recovery. Ich is a constant presence in the system due to the surface water supply, but tends to cause problems when fish are either stressed or predisposed by other infectious or noninfectious disease agents.

At the Lahotan NFH, Ich has been a problem about every other year in Lahotan cutthroat trout. Formaldehyde treatments at 153 ppm for 40 min, repeated three times, controlled the parasite. This treatment did not have a negative effect on the biofilter. After each production cycle, the whole system is disinfected with 200 ppm chlorine. This facility has a low incidence of diseases due to the low stocking densities, the use of well water, the high percentage of fresh water exchanged, and the complete disinfection between production cycles.

Ichtyobodo

Ichtyobodo is a small protozoan which infests the skin and gills. Infested fish may stop feeding and be listless or flashing may occur if skin is affected. A bluish film may develop over the body in heavily infested fish. This parasite was reported in one facility, where control became a problem. Treatment with formalin or salt did not control the infestation, even though the whole system, including the biofilter, was treated. A treatment regimen consisting of two treatments with 0.25 ppm copper sulfate for 24 hr at three day intervals and a third one at 0.50 ppm, plus a 30 day continuous treatment with 0.30-0.38 ppm controlled the infestation and did not adversely affect the biotilter. Water alkalinity was 45-55 ppm.

Proliferative Kidney Disease

Proliferative kidney disease (PKD) is an interstitial hyperplasic nephritis caused by an unclassified myxosporean parasite in salmonids. PKD has been a major problem among farmraised rainbow trout in Europe, but it has also been reported in the western United States and Canada. The most commonly infected species is the rainbow trout. Most epizootics of PKD occur at temperatures 2 15 °C. External clinical signs include darkening body color, distended abdomen due to ascites, pale gills, pronounced lateral body swelling, and bilateral exopthalmia. Internal clinical signs include enlargement of the kidney and spleen. The kidney may be grayish throughout or mottled and markedly swollen; in severe cases, the capsule may have a folded or corrugated appearance. Ascites may be present and is usually clear.

PKD was reported in rainbow trout cultured in an indoor recirculating unit of the Fish Experimental Facilities of the Research Centre of Jouy-en-Josas, France. The outbreak resulted in a 30% mortality of 17-month old rainbow trout. The average water temperature had increased from 14 to 18.5 °C during the two months prior to the mortalities. The fish had been hatched and raised at the facility. The unit also held in separate tanks other adult trout species, common carp, and goldfish (C'arussius aurutus). Evidence of myxosporean parasite shedding was found in the goldfish, which may have been the source of infection. Experiments suggested that indirect transmission of PKD via sediments was possible.

Schlotfeldt reported an outbreak of PKD in a recirculating unit in Germany. He suggested that introduction of infected rainbow trout was responsible for contamination of the system. The water temperature was maintained between 15-l 8 °C, which resembles the summer water temperatures present when natural outbreaks occur. These studies suggest that an increase in water temperature may precipitate an outbreak of PKD, if infected fish have been introduced into the system or if fish are in contact with contaminated sediments.

Amoebic Gill Infestation

An unidentified amoeba infested the gills of rainbow trout held in a recirculating system at the Freshwater Institute. BGD occurred in five consecutive lots 6-8 days after stocking; fish were treated with Chloramine-T several times, after which the amoeba infested the gills. Fish infested with amoebae swam near the surface with flared opercula. Macroscopically, gills had white inflamed tips. Histological sections showed large numbers of amoebae (10 pm diameter) on the surface of the gills associated with severe hyperplasia. If left untreated, the amoeba caused chronic mortalities for periods up to 2 months. The mortalities may have been caused by asphyxiation due to large numbers of amoebae on gill surfaces. Mortalities were controlled by treating with 167 ppm formalin every other day, for three treatments.

The amoeba was probably introduced with the fish, as it was observed in one lot during quarantine. The amoeba may be an opportunistic microorganism that thrives in conditions present in the recirculating system. Because the amoeba was a secondary infestation, BGD may have provided favorable conditions for amoebic infestation. Amoeba may use bacteria, detritus, or mucus as a food source. Because consecutive lots of fish were kept in the same tank, the amoeba may have become established in the population and/or in the biofilter. In European catfish, raised in a heated partial recirculating system, cells resembling amoeba trophozoites were associated with severe proliferative gill disease and systemic granulamatous tissue reactions. Mortalities reached 30% from this normally free-living amoeba, which multiplied when water quality was suboptimal and there was a heavy bacterial load.

Coleps

Coleps sp., a freshwater ciliated protozoan not normally considered pathogenic, was associated with 5-10 mortalities per day of 15 to 18 cm rainbow trout maintained in a densely stocked water recirculating system in central New York State. The protozoan was found in wet mounts of skin scrapings and gill tissue. Moribund fish showed excess mucous production. Histologically, severe multifocal interlamellar tissue proliferation was observed on the gills, and multifocal areas in the skin had eroded or missing epidermis. High numbers of mucous cells were observed in the skin and gills.

Wooster and Bowser reported that microscopic examination of water samples showed a bloom of Coleps sp. with a density estimated to be 7,000/ml. When the infestation appeared, nitrite level was high and salt was routinely added to maintain a chloride ion concentration of 190 mg/L. Treatment with formalin at a concentration of 20 mg/L for 5 days and at 25 mg/L for an additional

5 days proved ineffective. Treatment with an additional 0.15% sodium chloride did not reduce Coleps sp. counts in the water after 48 hr. Total mortality reached 35%. Potassium permanganate at 25 mg/L destroyed Coleps sp. in laboratory bioassays, but it could not be used, as that concentration would be toxic to the biolilter. Mortalities may have been due to excessive skin irritation caused by overwhelming numbers of protozoans. Conditions within the recirculating system probably provided an ideal environment for Coleps sp. to multiply and thus become a problem.

Saprolegnia

An external fungal infection caused by Saprolegnia sp. is often considered a secondary infestation following injury. As fish grow, the lesions enlarge and may eventually cause death. Saprolegnia sp. caused problems in recirculating systems in Canada when square or rectangular tanks were used to culture rainbow trout or Atlantic salmon (Salmo safar). These tanks allow feed and feces to accumulate, which promotes fungal development. However, treatment options were limited due to experiments that were being conducted. Prior to initiation of treatment, severely afflicted fish were removed from the tanks, and food or feces removed with a fine mesh dip net. The treatment consisted of a bath in saltwater (artificial seawatersalt) at a concentration of 1 to 1.5% for 1 hr. Dissolved oxygen and temperature were monitored during treatment, and fish were monitored for any signs of distress. After treatment, water flow to the tank was increased to flush the saltwater. If the problem recurred in a specific tank, fish were removed and treated again in a separate tank. The infected tank was drained, scrubbed, and disinfected with 200 mg/L sodium hypochlorite solution. After thorough rinsing with fresh water, the tank was refilled and the fish returned. It is necessary to improve the sanitary conditions in which fish are cultured to prevent Saprolegnia sp. from becoming a problem.

Infectious Diseases-viral

There are few references on viral diseases of rainbow trout in recirculating systems.

Infectious Pancreatic Necrosis

Infectious Pancreatic Necrosis (IPN) virus and other Bimaviruses are distributed worldwide. In salmonids, acute infection occurs in 1 to 4 month-old fish and can result in mortalities approaching 100%. Fish 26 months old undergo subclinical or inapparent infection, but few die. Disease outbreaks in older fish involve virus carriers and are usually stress activated. Since there is no effective treatment, IPN is most effectively controlled by preventing contact between the host and the virus. Fish introduced into a facility should be assayed and determined to be specific pathogen-free, and eggs should originate from virus-free broodstock because IPN virus appears to be transmitted with eggs in spite of iodophor treatment. The incidence of acute IPN and consequent mortality can be reduced if factors that cause stress are controlled: reducing population density, following optimal feeding protocols, maintaining proper hatchery hygiene, and utilizing prophylactic treatments for bacterial diseases and parasites have been effective in moderating outbreaks of IPN.

Munro and Fijan reported that IPN was considered to be a continuous presence and an intermittent problem in young rainbow trout in European culture systems using heated effluents and/or some form of recirculation. Survivors of epizootics tended to be poor growers.

Viral Hemorrhagic Septicemia

Viral hemorrhagic septicemia (VHS) is enzootic in most countries of continental eastern and western Europe, and in the United States the virus can be isolated in the Puget Sound area of Washington and in the Gulf of Alaska. In Europe, epizootics occur primarily in rainbow trout, brown trout (Salmo trutta), and in northern pike (Esox lucks). In rainbow trout, VHS passes through three stages: an acute stage with high mortalities, a chronic stage where fish appear dark and anemic, and the third stage where erratic swimming and nervous behavior is characteristic and when mortality is low. Cooler water temperatures (4.4-7.7"C) favor VHS outbreaks in rainbow trout, and stress factors such as crowding, handling, transportation, or malnutrition also predispose fish to VHS.

VHS is considered a hazard to each new year-class of rainbow trout stocked in European culture systems using heated effluents and/or some form or recirculation.

Infectious Hematopoietic Necrosis

Infectious hematopoietic necrosis (IHN) is endemic to the Pacific Coast of North America, and it is well established in Japan. The virus has been disseminated to other areas of the United States and the world via contaminated fish and fish eggs originating in the Pacific Northwest of North America. Epizootics of IHN can result in high mortality in 3-week to 6-month-old fish. Mortalities can occur in 7 to 14 month-old fish, but clinical signs seen in younger fish are generally absent. As there is no known treatment for the disease the most effective means for controlling IHN is by preventing contact between the virus and the host. Broodstock culling and disinfection of eggs with iodophor are partly effective, but epizootics can occur despite this management strategy. Stocks of eggs should originate from virus-free broodstock (to prevent vertical transmission), and stocks of young-of-the-year-fish should be assayed and determined to be BIN-free. Ideally, the water supply should be controlled and II-IN-free. Water can also be effectively disinfected using ozone or ultraviolet irradiation.

The IHN virus is endemic at Dworshak NFH, and epizootics have occurred in steelhead trout during the operation of the recirculating systems, but outbreaks are more frequent earlier in the culture program when fish are younger and the systems are operated with single-pass surface water. The recirculating systems may be returned to single-pass when IHN infections occur to reduce horizontal transmission of the virus and severity of infection.

Non-infectious Diseases

Temperature and certain chemical constituents in water can create non-infectious problems or be toxic. Non-toxic levels of contaminants can stress fish, which can lead to disease outbreaks or reduced growth. Recirculating aquaculture systems, in particular, can have poor water quality due to improper temperature control or performance of any of the treatment units. The most important environmental variables within recirculating systems that can cause problems are low levels of oxygen or high levels of un-ionized ammonia, nitrite, carbon dioxide, suspended solids, gas supersaturations, or ozone. The water's pH also plays a role, because acid-base chemistry controls the proportion of the total ammonia nitrogen and inorganic carbon that exist as un-ionized ammonia and carbon dioxide, respectively. Therefore, water quality needs to be monitored closely in a recirculating system so that problems with the water treatment units can be detected early and corrected.

Temperature

The optimum temperature for growth of rainbow trout is 15 °C. Maintaining the optimum water temperature is an important issue in recirculating systems. Water heats up from pumping and from heat transfer through the pipes and tank walls as it is retained and reused in a semiclosed environment. This temperature gain can be advantageous in instances when cold ground waters are brought up to 15 °C; however, a temperature gain much beyond this can have serious fish health consequences, including decreased growth rates and increased incidence of stress and disease.

At the Freshwater Institute, water within the recirculating rainbow trout-culture system is maintained at 13-16 °C by exchanging the total system water volume twice a day with 12.5 °C ground water; outside of water exchange, the system is not actively heated or chilled. The high water exchange rate has been used to keep water temperatures below 17 °C during summer months. In contrast, recirculating systems for culture of warm-water fish exchange relatively little water to reduce the amount that must be heated.

Ammonia and Nitrites

The most common non-infectious problem in recirculating systems may be high ammonia or nitrite levels, which are produced by biofilter malfunction. Nitrate is less likely to accumulate to toxic levels than either ammonia or nitrite within recirculating systems because it is much less toxic to fish; Braun reported that rainbow trout tolerated nitrate levels up to 800 mg/L. The 96-h LC50 value of nitrite for rainbow trout is 0.20 to 0.40 mg/L. Nitrite toxicity is pH-dependent and is also dependent, in varying degrees, on the presence of chloride, sulfate, phosphate, and nitrate ions. Wedemeyer and Yasutake reported that nitrite toxicity was reduced as pH increased. Russo and Thurston reported that the presence of chloride suppressed nitrite toxicity in rainbow trout, and Bowser et al. found chloride to have a protective effect against nitrite toxicity in Atlantic salmon (Safmo salar). Sodium chloride is used to control nitrite toxicity in fish culture systems, as the chloride is taken up by the gills preferentially to the nitrite. According to Colt and Armstrong, water containing calcium or chloride ions can increase the tolerance of rainbow trout to nitrite by a factor of 20 to 30. Even so, the desirable level of nitrite for rainbow trout culture in a reuse system is <0.1 mg/L.

Recirculating systems use biofilters to reduce dissolved wastes, particularly ammonia. Within the biofilter, heterotrophic micro-organisms and autotrophic bacteria biologically oxidize a portion of the ammonia nitrogen and the biodegradable organics. The bacterial conversion of ammonia to nitrate is a two-step process called nitrification. Two different groups of nitrifying bacteria, both believed to be obligate autotrophs and obligate aerobes are needed for complete nitrification. Nitrosomonas convert ammonia to nitrite and Nitrobacter converts nitrite into nitrate. To avoid accumulations of nitrite and ammonia, time must be provided to allow bacterial populations to become established within the biofilter when biofilters are started up or when the organic, ammonia, or nitrite load to the biofilter are signficantly increased. Accumulation of ammonia can occur in an initial lag period during biofilter start-up because the Nitrosomonas population has not become established; accumulation of nitrite can occur in a second lag period because the nitrobacter population has not become established.

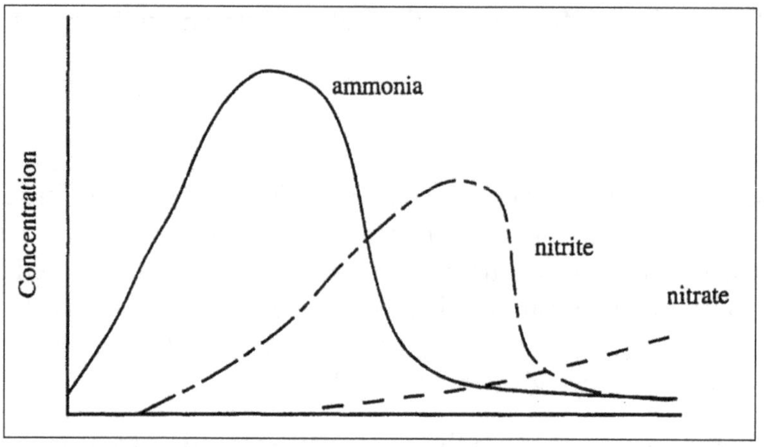

Temporal sequence and relative magnitude of the accumulation of ammonia, nitrite, and nitrate concentrations as a new biofilter acclimates to the substrate loading in a recirculating system.

At the Freshwater Institute, the biofilter microflora was developed naturally by stocking the tanks at a low density and by allowing the ammonia produced by the fish to increase the nitrifying bacteria population. Bacteria multiply more slowly at the lower water temperatures found in the Freshwater Institute system (e.g., 15 °C), so there may have been even longer lag periods when ammonia and then nitrites accumulated. The problem disappeared in about 8 weeks, once the biofilter reached a steady state. At the Freshwater Institute, when nitrite reached a level of 0.3 mg/L, sodium chloride was added directly to the culture tanks at 10 times the nitrite concentration to reduce toxicity. Abrupt increases in salinity may impair nitrification and biofilter recovery.

Elevated nitrite levels also occurred on several occasions that may be attributed to biofilter defluidization, exposure of the biofilter to chemotherapeutants, or huge and rapid increases in fish/feed loading on the biofilter. In one particular example, during a biofilm stripper study, too much biofilm was sheared from the biofilter. Nitrite levels increased and the ensuing situation was similar to that which occurs when the microflora in a new biofilter is developed, with a lag in the Nitrubacter population. In this case, the levels of nitrite reached 0.8 mg/L, and clinical signs of methemoglobinemia were present in the fish. The fish were lethargic and had flared gills that were brownish in color. Some fish with clinical signs of furunculosis also were detected. Nitrite-induced mortality has been rare at the Freshwater Institute, but it appears that high nitrite levels act as a stressor that has precipitated outbreaks of infectious diseases.

Another example of high nitrite levels occurred at Freshwater Institute during system ozonation. Ozone stoichiometrically oxidizes nitrite to nitrate; therefore, addition of ozone to recirculating systems is beneficial because it reduces nitrite levels. The fact that ozone decreases the levels of nitrite is a substantial benefit on those occasions when bacterial conversion of nitrite to nitrate in the biofilter is lost. However, because ozone actually reduces the nitrite concentration going to the biofilter, it also reduces the quantity of bacteria converting nitrite to nitrate and thus reduces the total acclimated nitrite removal capacity of the biofilter. The drawback to ozonation occurs when addition of ozone is interrupted. Because ozone is responsible for removing a fairly large fraction of the total nitrite produced in the recirculating system, when ozone addition is interrupted, nitrite rapidly accumulates within these systems.

Carbon Dioxide

Carbon dioxide is toxic to fish at levels of 20-40 mg/L because it significantly reduces the capacity of blood to transport oxygen, and it can produce nephrocalcinosis. Carbon dioxide can accumulate to toxic levels in a recirculating system with pure oxygen addition, due to low water gas exchange and high fish loading. The level of dissolved carbon dioxide is dependent on pH, alkalinity, and temperature. If the water has been aerated sufficiently to come to equilibrium with the atmospheric partial pressure of carbon dioxide, it will contain about 2 mg/L of dissolved carbon dioxide. In fact, aeration and air stripping are important mechanisms for removing accumulations of carbon dioxide in intensive recirculating systems.

Fish excrete carbon dioxide through their gills. Carbon dioxide unloading is a function of both ventilation rate and concentration gradient. If levels of carbon dioxide in the water are elevated, less carbon dioxide can be transferred from the gills into the water. Rainbow trout control the rate water is pumped through their gills based upon the blood's oxygen content. High carbon dioxide levels in the water result in a reduction of blood oxygen in the fish. Based upon an extensive literature review, Heinen et al. estimated that the upper safe levels for chronic exposure to carbon dioxide in rainbow trout range from <9-30 mg/L. The 20 mg/L recommended safe level may be conservative, especially if oxygen in the water is high. Due to buffering and acid-base chemistry, water with high alkalinity or pH or both may enable fish to tolerate higher concentrations of free carbon dioxide. This may account for the varying levels reported as safe in the literature.

Water pH and alkalinity play an important role in determining the amount of carbon dioxide present. According to acid-base equilibrium, the amount of carbon dioxide increases as the pH decreases. This also means that in highly alkaline water, toxic levels of carbon dioxide occur at higher pH levels than in low-alkalinity water. When carbon dioxide was first identified as a fish health problem in the recirculating system at the Freshwater Institute, there was insufficient contact between air and water to strip out the carbon dioxide produced by fish metabolism. In the first year of operation, the high fish loadings (3.0 KgL/min) and inadequate air-to-water contact produced accumulations of carbon dioxide that dropped the pH to about 7.0 and translated into a carbon dioxide concentration of about 50 mg/L. When fish were exposed to these levels of carbon dioxide (50 mg/L), they appeared initially agitated, exhibited jumping behavior, and then became listless and sluggish, as in anoxia, before dying. Moribund fish had gaping mouths and flared operculums and their gills were a brilliant cherry red color. Beyond a certain point, moribund fish did not recover when placed in fresh water. Mortalities were nearly 100%. In small-scale tests conducted at that time, rainbow trout mortality from carbon dioxide toxicity occurred at pH levels 56.8 and alkalinities of about 250 mg/L (as calcium carbonate); these conditions translate into carbon dioxide concentrations 2 79 mgL.

Heinen et al. reported that modifications to the hybrid oxygen contactor/stripping column used at the time produced more air to water contact, which reduced carbon dioxide to about 17-32 mg/L, at an alkalinity of 215-276 mg/L (as calcium carbonate) and increased pH to 7.2-7.4. The system was again modified in 1992 with the installation of a stripping tower, designed according to criteria reported by Summerfelt. Subsequently, carbon dioxide concentrations have averaged 16 mg/L.

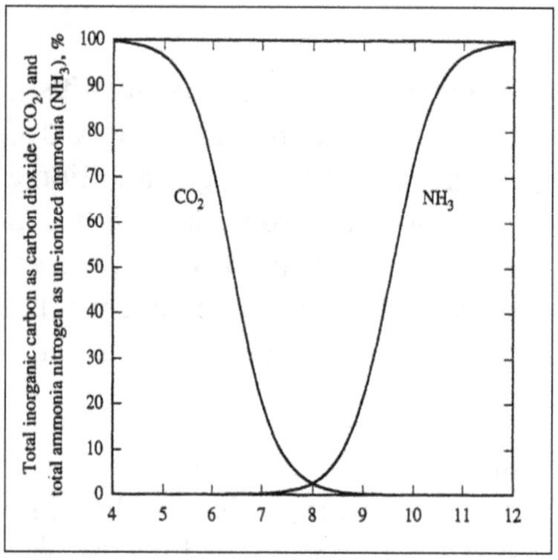

Percent total inorganic carbon as carbon dioxide (CO_2) and the percent total ammonia nitrogen as free ammonia (NH_3) as a function of pH, assuming equilibrium at 15 °C in freshwater.

pH

Water pH plays an important role in aquaculture, because it controls the equilibrium of the ammonia and the carbonic acid systems. Un-ionized ammonia (NH_3) is the toxic form of ammonia; it associates with water to form hydroxide (OH^-) and ammonium (NH_4^+). Dissolved carbon dioxide is the toxic form of the inorganic carbon system, which includes carbonic acid (H_2CO_3), bicarbonate (HCO_3^-), and carbonate (CO^{2-}).

Comparing the range of pH values where carbon dioxide and ammonia coexist indicates that the smallest fractions of each coexists at a pH of 8.0, assuming the water is 15 °C. The equilibrium carbon dioxide or ammonia concentration changes 10 fold for every 1 unit change in pH.

Adding a source of alkalinity (e.g., lime, caustic soda, soda ash, or sodium bicarbonate) to the water in a recirculating aquaculture system can help to maintain a pH that will minimize the potentially toxic effects of carbon dioxide and ammonia.

Ozone

Roselund reported gill epithelial damage and death of rainbow trout exposed to an estimated 0.010-0.060 mg/L of ozone. Wedemeyer et al. reported a lethal threshold level of about 8 I.tg/L, with death apparently resulting from massive destruction of the gill lamellar epithelium and severe hydromineral imbalance. The maximum safe level of chronic exposure for salmonids was provisionally set at 2 pg/L. Oakes et al. found that most steelhead trout exposed to ozone concentrations 2 0.20 mg/L died within 4 hr, while those exposed to 0.008 mg/L did not die during 50 hr of exposure.

During research on ozonation at the Freshwater Institute, ozone was added to the water with the oxygen feed gas just before it entered the culture tanks at a rate of 0.025-0.039 kg ozone per kilogram of feed fed. Mortality caused by ozone residual in the tanks occurred on five occasions causing 3.9% and 5.0% mortalities in two lots of rainbow trout. Oxidation-reduction potential (ORP) probes and controllers were used to limit ozone toxicity; however, tank locations away from the probes accumulated toxic concentrations. The first clinical signs of exposure to toxic concentrations of dissolved ozone were noticeable changes in fish behavior. Fish stopped feeding and congregated near the surface and sometimes "gasped" for air. Eventually, erratic swimming behavior occurred and became progressively worse. Attempts to jump out of the tank increased, and some fish showed darting behavior followed by listless swimming. Fish then lost equilibrium and became pale, with vertical patches of dark pigment on the sides of the body. Fish which reached this latter condition rarely survived. Gills of fish exposed to high levels of ozone showed excess mucus, hyperplasia, and aneurysms.

Solids

Suspended solids have been associated with environmentally-induced disease problems; in certain intensive systems, suspended solids have been cited as one of the major factors limiting production. Solids have been reported to cause sublethal effects such as fin rot and direct gill damage in rainbow trout. At the Freshwater Institute, Bullock et al. reported that suspended solids within the recirculating system were thought to contribute to BGD outbreaks in rainbow trout.

Solids are produced in recirculating systems as uneaten feed, feed fines, fish fecal matter, and sloughed biofilter cell mass. Rainbow trout feces are contained within a mucous sheath that can remain intact if they are removed soon after deposition. If they are not quickly removed, shear forces break the mucous sheath, which allows the fecal matter to disintegrate into much finer and more soluble particles. Due to the detrimental effects of solids on both fish health and the performance of the other components within the recirculating system, solids removal is one of the most critical process in managing recirculating aquaculture systems.

Gas Supersaturation

Gas supersaturation can occur when gas comes into contact with water at pressures greater than atmospheric, or when water is heated and the saturation concentration of gas decreases. Often not considered is gas supersaturation produced when air is aspirated into water flowing at a high velocity through leaky pipes, even under pressure. Gas supersaturation can also arise from passing air into deep water. It is the nitrogen and argon component of ambient air that are the most detrimental component of total gas saturation. Nitrogen saturation levels greater than 110% affect juvenile fish and levels greater than 125% affect adults. However, problems with high dissolved gas pressures are not generally encountered in a properly designed recirculating system.

Prevention and Control

The same accepted principles for disease prevention and control apply to rainbow trout raised in recirculating systems as to those reared in flow-through systems. Most of these management practices become more critical in a recirculating system because fish are held at much higher densities; the system has added components, such as the biofilter which may harbor disease organisms; and carrier fish can reside in the system even after recovery from an epizootic which is critical if continous production strategies are used.

The main preventative measure is to keep known pathogens from entering the system, by avoiding introduction of infected fish. This can be accomplished by purchasing eggs from diseasefree broodstock. If this is not possible, fingerlings should be bought from a disease-free certified source, quarantined upon arrival, and checked for parasites and bacteria before they are stocked. The water should be spring or well water, because these sources are usually pathogen-free, or if surface water is used, it should be disinfected before entering the system.

If a disease enters a recirculating system and an outbreak occurs, the organism can spread rapidly throughout the population due to the high loading density and to the recirculation of water. Avoidance of cross contamination between tanks could contain an outbreak to the affected tanks as long as the water treatment units are independent. Equipment entering the facility and equipment used in production should be regularly disinfected. Each tank should have its own nets and other equipment.

The key to success once an outbreak occurs is prompt diagnosis and treatment. Routine observation of mortalities and abnormal swimming or feeding behavior is useful in early detection of a disease problem. It is important, at the same time, to reduce or to eliminate any stressor which may have contributed to the disease outbreak. For example, water quality may act as a stressor, if the ammonia, nitrites, and/or suspended solids are high.

If a disease is resistant to treatment, it may be necessary to eliminate the infected population and disinfect the tanks. When a continuous production method is used, where fish of different cohorts are in the same tank, a disease organism may establish itself in the population and then infect newly introduced fingerlings.

In Europe, experience with small, intensive systems suggests that they can be run with minimal health problems. However, it appears that commercial systems have periods of good performance followed by periods of indifferent performance and often disease in the populations. Only sometimes can a cause of poor performance or disease be found. Due to the artificial nature of recirculating system environments, any number of opportunistic microorganisms, if given a favorable environment, could reproduce, establish themselves in the system and possibly become pathogenic. Diseases could become a limiting factor in commercial systems, due to the possibility that a disease organism may become established in the population or in the biofilter and spread to uninfected fish in the system. Economic losses sustained because of diseases are those due to mortalities, deformities, and lesions; poor growth rates; and poor food conversion of the survivors of a disease outbreak.

Each recirculating culture facility should design a protocol for prevention of and control of fish diseases with the aid of a fish health professional, based on the generally accepted principles of fish health management.

Mussel

Mussels are among the many invertebrates under the Phylum Mollusca. Their wide distribution in the coastal areas of the Indo-Pacific region makes them the most easily gathered seafood organisms, contributing a significant percentage to the world marine bivalve production. In the Philippines, approximately 12,000 MT of mussels were produced in 1987. This amount consisted only of farmed green mussel, Perna viridis, and not the brown mussels which are exclusively gathered from natural beds.

In the wild, mussels are mostly found in the littoral zone, attached in clusters on various substrates. Being a filter-feeder of phytoplankton and detritus, it is considered the most efficient converter of nutrients and organic matter, produced by marine organisms in the aquatic environment, into palatable and nutritious animal protein. Its very short food chain, sturdy nature, fast growth rate and rare occurrence of catastrophic mass mortalities caused by parasitic micro-organisms, makes it possible to produce large quantities at a very reasonable price. Likewise, its ability to attach to substrates with the byssus, makes it an ideal aquaculture species using different culture systems. According to Bardach et al., mussel culture is the most productive form of saltwater aquaculture and its proliferation is virtually a certainty.

France can probably be credited to have the longest history of mussel culture which dates as far back as 1235, while Spain has been reported to be the top world producer of farmed mussels.

In the Philippines, mussel culture started only in 1962 at the Binakayan Demonstration Oyster Farm, in Binakayan, Cavite by the biologists of the then Philippine Fisheries Commission, now Bureau of Fisheries and Aquatic Resources (BFAR). Mussels were initially considered as a fouling organism by oyster growers. The impetus for mussel culture in Manila Bay came about when oyster growers, attempting to collect oyster spats in less silty offshore waters, obtained instead exceptional heavy and almost pure mussel seedlings.

Mussel farming does not require highly sophisticated techniques compared to other aquaculture technologies. Even un-skilled laborers, men, women, and minors can be employed in the preparation of spat collectors as well as harvesting. Locally available materials can be used, hence minimum capital investment is required. The mussel harvest can be marketed locally and with good prospects for export.

Success in mussel farming, however, depends in providing some basic requirements to the bivalve such as: reasonable amount of sheltering of the culture areas, good seawater quality, and sufficient food in the form of planktonic organisms. These pre-requisites are found in some coastal waters, hence locating ideal sites for mussel cultivation is essential.

Biology of the Mussel

The green mussel, Perna viridis has separate sexes, although hermaphrodism usually occurs. Externally, it would be difficult to determine the sex, however, internally, the gonad tissue of a sexually matured male appears creamy-white in color, while that of the female is reddish-apricot. Sometimes young sexually immature females can not be distinguishable by color from male specimens.

This bivalve species reaches sexual maturity within the first year and spawns with the rising of seawater temperature. In the Philippines, mussels spawns year-round, however the peak period of spawning and setting is in April and May and again in September to October. Eggs and sperms are shed separately and fertilization occurs in the water.

Mussels have two relatively distinct phases in their life-cycle. A free swimming planktonic or larval stage and a sessile adult stage. The free swimming larvae remains planktonic for 7–15 days depending upon the water temperature, food supply and availability of settling materials. At about 2–5 weeks old, the larvae (0.25–0.3 mm) seek a suitable substrate to settle on and final metamorphosis takes place, changing its internal organ structure to the adult form. The young spat then grow rapidly and within 4–8 weeks, after settlement, they measure 3–4 mm in shell length.

Subsequent growth of the bivalve can be distinguished into shell and body growth. The shell length does not necessarily reflect the meat content. During spawning or food shortage, internal energy reserves are consumed while the shell may continue to grow. Overall growth of the mussel, as far as shell measurement is concerned is influenced by factors like temperature, salinity, food availability, disturbances and competition for space. On the other hand, body growth is affected by the season which primarily relate to sexual cycle and over-crowding to a certain extent.

Criteria for Site Selection

Site Location

In prospecting sites for mussel cultivation, well-protected or sheltered coves and bays are preferred than open un-protected areas. Sites affected by strong wind and big waves could damage the stock and culture materials and, therefore, must be avoided. Another important consideration is the presence of natural mussel spatfall.

Areas serving as catchment basins for excessive flood waters, during heavy rains, should not be selected. Flood waters would instantly change the temperature and salinity of the seawater, which is detrimental to the mussel.

Sites accessible by land or water transportation are preferred so that culture materials and harvests can be transported easily.

Water Quality

Areas rich in plankton, usually greenish in color, should be selected. Water should be clean and free from pollution. Sites near densely populated areas should not be selected in order to avoid domestic pollution. In addition, the culture areas should be far from dumping activities of industrial wastes and agricultural pesticides and herbicides.

Waters too rich in nutrients, which may cause dinoflagellate blooms and render the mussels temporarily dangerous for human consumption, causing either gastro-intestinal troubles or sometimes paralytic poisoning, should be avoided.

Water physio-chemical parameters are also important factors to be considered. The area selected should have a water temperature ranging from 27–30 °C, which is the optimum range required for mussel growth. Water salinity of 27–35 ppt is ideal. A water current of 17–25 cm per second during flood tide and 25–35 cm per second at ebb-tide should be observed.

Favourable water depth for culture is 2 m and above, both for spat collection and cultivation.

Bottom Type

Bottom consisting of a mixture of sand and mud has been observed to give better yields of mussel than firm ones. It also provides less effort in driving the stakes into the bottom. Shifting bottoms must be avoided.

Cultured Mussel Species

Among the mussels proliferating in the coastal areas of the tropical zone, the green mussel, Perna viridis (= Mytilus smaragdinus), called tahong in the Philippines, is the only species farmed commercially. In the temperate zone, it is the blue mussel, Mytilus edulis, as this species can grow at low seawater temperatures.

The brown mussel, Modiolus metcalfei and M. philippinarum which form dense mats on muddy bottoms in shallow bays are simply gathered.

Culture Methods

Mussel culture, as practiced in many countries, is carried out by using a variety of culture methods based on the prevailing hydrographical, social and economic conditions.

Bottom Culture

Bottom culture as the name implies is growing mussels directly on the bottom. In this culture system a firm bottom is required with adequate tidal flow to prevent silt deposition, removal of excreta, and to provide sufficient oxygen for the cultured animals.

Mussel bottom culture is extensively practiced in The Netherlands, where the production of seeds is completely left to nature. If the natural spatfall grounds are unsatisfactory for growing, the seedlings are transferred by the farmer to safer and richer ground or to his private growing plots, until the marketable size is attained. Natural conditions control the quality and quantity of food in the water flowing over the farming plots. Marketable mussels are fished from the plots and undergo cleansing before being sold.

This method requires a minimum investment. Disadvantages, however, of this type of culture is the heavy predation by oyster drills, starfish, crabs, etc. Also, siltation, poor growth and relatively low yields per unit culture area.

Mussel bottom culture.

Intertidal and Shallow Water Culture

The culture methods that fall under this category are usually practiced in the intertidal zone. The culture facilities are set in such a way that the mussels are submerged at all times. Culture methods are:

Rack Culture

This is an off-bottom type of mussel culture. Rack culture is predominantly practiced in the Philippines and Italy where sea bottom is usually soft and muddy, and tidal range is narrow. The process involves setting of artificial collectors on poles or horizontal structures built over or near natural spawning grounds of the shellfish. In the Philippines, this is called the hanging method of mussel farming. The different variations used are as follows:

1. Hanging method: The process starts with the preparation of the spat collectors or cultches. Nylon ropes or strings, No. 4, are threaded with coco fibre supported by bamboo pegs or empty oyster shells at 10 cm intervals. These collectors are hung on horizontal bamboo poles at 0.5 m apart. A piece of steel or stone is attached at the end of the rope to prevent the collector to float to the surface. Setting of collectors is timed with the spawning season of the mussels. Spats collected are allowed to grow on the collectors until marketable size.

Other materials utilized as collectors are rubber sheets and strips from old tires.

Mussels are harvested by taking out from the water the ropes or strings and bringing them to the shore on a banca. The same collectors can be re-used after being cleaned of fouling organisms. Harvested mussels are cleansed of the dirt and mud by dipping the collectors several times in the water. The process maybe laborious, but the ease in harvesting and availability of local materials for culture purposes makes it very adaptable under local conditions.

Mussel hanging (bitin) culture method on bamboo plots as practiced in the Philippines.

Mussel stick (tulus) culture method as practiced in the Philippines.

2. Stake (tulos) method: The stake method is midway between the rack and bottom methods. Bamboo poles, 4–6 m in length are staked firmly at the bottom in rows, 0.5–1 m apart during low tide in areas about 3.0 m deep and above. In areas where water current is strong, bamboo poles are kept in place by nailing long horizontal bamboo supports between rows. Since mussels need to be submerged at all times, it is not necessary that the tip of the poles protrude above the low water

level after staking. However, boundary poles should extend above the high water level. In staking, enough space between plots is allowed for the passage of the farmer's banca during maintenance.

Collected spats are allowed to grow in-situ until marketable size, 5–10 cm after 6–10 months. It has been observed, that about 2,000–3,000 seeds attach on 1 metre of stake, 1–2 m below low water level.

The mussels are harvested by pulling out the poles and bringing them ashore on a banca. Some poles may still be sturdy and can be re-used during the next season.

3. Tray culture: Tray culture of mussels is limited to detached clusters of mussels. Bamboo or metal trays, 1.5 m × 1 m × 15 cm sidings are used. The tray is either hang between poles of the hanging or stake methods or suspended on four bamboo posts.

4. Wig-wam culture: The wig-wam method requires a central bamboo pole serving as the pivot from which 8 full-length bamboo poles are made to radiate by firmly staking the butt ends into the bottom and nailing the ends to the central pole, in a wigwam fashion. The stakes are driven 1.5 m apart and 2 m away from the pivot. To further support the structure, horizontal bamboo braces are nailed to the outside frame above the low tide mark. Spats settle on the bamboos and are allowed to grow to the marketable size in 8–10 months.

Mussels are harvested by taking the poles out of water, or in cases that there are plenty of undersized bivalves, marketable mussels are detached by divers.

5. Rope-web culture: The rope-web method of mussel culture was first tried in Sapian Bay, Capiz, in 1975 by a private company. It is an expensive type of culture utilizing synthetic nylon ropes, 12 mm in diameter. The ropes are made into webs tied vertically to bamboo poles. A web consists of two parallel ropes with a length of 5 m each and positioned 2 m apart. They are connected to each other by a 40 m long rope tied or fastened in a zigzag fashion at an interval of 40 cm between knots along each of the parallel ropes. Bamboo pegs, 20 cm in length and 1 cm width are inserted into the rope at 40 cm interval to prevent sliding of the crop as it grows bigger.

In harvesting, the rope webs are untied and the clusters of mussels are detached.

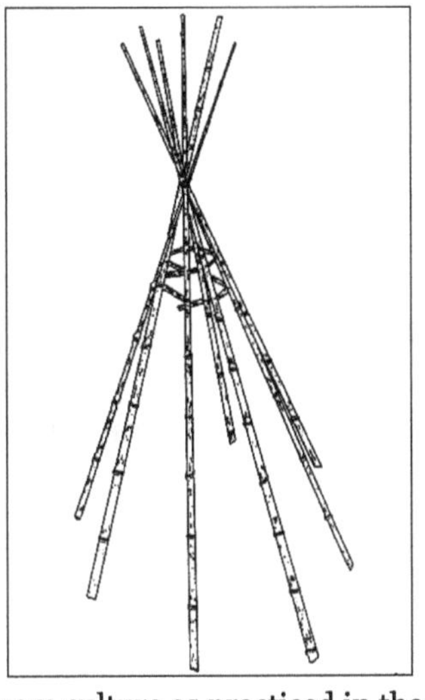

Mussel wig-wam culture as practiced in the Philippines.

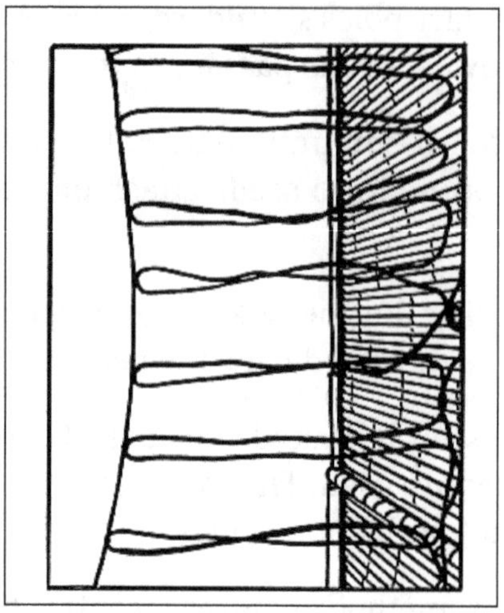

Mussel rope-web culture method as practiced in the Philippines.

The method is laborious and expensive, but the durability of the ropes which could last for several years might render it economical on the long run. However, the effect of the culture method on the culture ground is detrimental as gradual shallowing of the culture area has been observed up to the point that the areas become no longer suitable for mussel farming.

"Bouchot" Culture

"Bouchot" culture is mainly undertaken in France. This is also called the "pole culture" or stake culture. The poles, used are big branches or trunks of oak tree, 4–6 m in length, which are staked in rows, 0.7 m apart on soft and muddy bottoms of the intertidal zone during low tide.

Mussel seeds are collected on coco-fibre ropes which are stretched out horizontally on poles. Young adults, 3–5 mm in size are placed in long netlon tubes (10 m in length) and attached around the oak poles in a spiral fashion, until marketable size.

Korringa reported that for an estimated length of about 600 km "bouchot" netlon, an approximate production of 7000 tons of marketable mussels yearly or an average production of 25 kg/pole/year can be harvested.

Deep Water Culture

Raft Culture

Mussel raft culture has been practiced in Spain for a long time. Mussel seeds that settle freely on rocks or on rope collectors are suspended from a raft. When the weight of the bivalves on a given rope exceeds a certain limit, the rope is taken out and again distributed over a greater length until marketable size. It is a continuous thinning of the mussel stock to provide ample space to grow. Marketable shellfish are detached from the rope, purified in basins before marketing.

The raft may be an old wooden boat with a system of outrigger built around it. Other kinds of rafts could be a catamaran-type boat carrying some 1000 rope hangings, or just an ordinary plain wooden raft with floats and anchors. Floats can be made of plastic, wood, oil drums, etc. The raft are transferred from one place to another using a motor boat.

Production of mussels from this type of culture is high. From a catamaran-type raft with 1,000 rope, 6–9 m in length, about 4,666–5,333 MT of marketable mussel can be produced.

Mussel raft culture method.

Advantages of this type of culture are: reduce predation, utilization of planktonic food at all levels of water, and minimum siltation.

Long-line Culture

Long-line culture is an alternative to raft culture in areas less protected from wave action. A long-line supported by a series of small floats joined by a cable or chain and anchored at the bottom on both end is employed. Collected mussel spats on ropes or strings are suspended on the line. The structure is fairly flexible.

Mussel Transplantation to New Site

Transplantation of young mussels from natural spawning grounds to sites with favourable conditions for growth is practiced in numerous countries as mentioned earlier. In the Philippines, however, mussel transplantation to new sites is being encouraged to develop new areas for mussel culture, due to various reasons. Major reasons are: rampant pollution of some existing mussel areas, urbanization growth near mussel farms and competitive use of lands.

Mussels to be transplanted could be breeders or young adults. Important points to be considered are: Conditions from natural spawning areas must be almost similar to the new area, mussels on original collectors showed better survival than those detached, and in transporting the mussel avoid being exposed to heat and freshwater.

Harvesting Procedure

Harvesters should be aware of the stress caused during the harvesting process. In harvesting mussels special care is needed. Pulling them or using a dull scraper may tear the byssal thread. This will result in loss of moisture after harvest or cause physical damage causing early death of the bivalve. The right procedure is to cut the byssal thread and leave it intact to the body. Exposure to sun, bagging and transport also increases the stress of the mussels.

Depuration of Mussels

To date, depuration of mussels in the Philippines is not yet undertaken due to its prohibitive cost. Mussel farmers cleansed their harvest by relaying them in clean water. This procedure, however, is unlikely to reduce heavy contamination by toxic wastes, accumulated during growing period.

Disease Management in Mussel

An important step in comprehending freshwater mussel health status is to gain information about pathogens. Parasites and infectious agents with related lesions (i.e., inflammations and regressive phenomena) of this group is poorly described in literature. Among the reported pathogens, bacteria, protozoan and metazoan parasites like trematodes, nematodes, mites, and ciliates (Conchophthirus spp.), have the potential to decrease the fitness of the host unionid, but their role in diseases has not been well established.

Regnum	Phylum	Class	Species	Bivalve hosts
Virus	Arenavirus		Lea plague Virus (HcPV)	Hyriopsis cumingii
Fungi	Heterokonta	Oomycota	Oomycetes saprofites	Unio spp.
Protozoa	Ciliophora		Conchophthirus spp.	Elliptio complanata, Anodonta marginata, Anodonta implicata, Pyganodon cataracta, Lampsilis radiata, Lampsilis cariosa, Alasmidonta undulata, Anodonta cygnea
			Heterocinetopsis unionidarum	Pyganodon (= Anodonta) grandis, Lasmigona complanata
			Trichodina unionis	Anodonta cygnea, Unio spp.
Metazoa	Platelmintes	Trematodes Digenea	Aspidogaster conchicola	Indonaia caerulea, Corbicula striatella, Lamellidens corrianus
			Cotylaspis insignis	
			Cotylogaster occidentalis	
			Lophotaspis interiora	
			Bucephalus polymorphus	Unio pictorum, Dreissena spp.
			Rhipidocotyle spp.	Unio pictorum, A. anatina
			Polylekithum spp.	A. plicata
		Nematoda	Hysterothylacium spp.	Diplodon suavidicus
	Artropoda	Copepods	Paraergasilus rylovi	Anodonta piscinalis
		Mites	Unionicola spp.	Unio complanata, Unio gibbosus U. ligamentinus U. intermedia, A. fragilis, A. footiana, A. cataracta, Anodonta cygnea, A. anatina, Elliptio complanata
Najadicola spp.				

Evidence for viral diseases has been found in only one species of freshwater bivalve, a Chinese pearl mussel, Hyriopsis cumingii. The Hyriopsis cumingii, Lea virus disease, which is often referred to Lea plague disease (HCPD), was first reported in 1980s. Next reports used light and transmission electron microscopic (TEM) analysis of tissues from diseased bivalve mussels showed that the HCPD was associated with an arenavirus agent termed Hyriopsis cumingii Lea plague Virus (HcPV).

Differently to marine bivalves, little is known about bacterial diseases of this group of molluscs. All the reports present in literature still are uncertain about their role as pathogens/symbionts. In general, the importance of bacteria as etiological agent of diseases in marine bivalves is mostly reported in intensively cultured species. The presence of different species of bacteria in the connective tissue and in the digestive gland of affected mussels, but related to scarce haemocytosis.

About protozoan reports, in unionids the most common group reported belong to the genus Conchophthirusspp. (family Conchophthiridae). Species of this genus are only found in freshwater bivalves, and are among the most common organisms in this animal group. The body of these ciliates is flattened, elliptical in profile, with the mouth near the middle of the body. They have dense cilia over their entire surface and an average length of about 100 μm. Conchophthirusspp. move within the mantle cavity and are not firmly attached to the host. The reported species by Kelly are Conchophthirus anodontae and Conchophthirus curtus were of 30 of the 44 species of unionids examined from Illinois and Pennsylvania.

Antipa and Small described the presence of the ciliate Heterocinetopsis unionidarum (Ancistrocomidae) in 2 of the 4 species of mussels examined in the one locality where it occurred. Parasitized unionids like Anodonta grandis and Lasmigona complanata didn't show specific harmful effects from this protozoa, present at the gills and palps. Other reported ciliates belong to the genus Tricodina. This genus and related genera (Peritrichia: Trichodinidae) include numerous species reported as fish and marine bivalve parasites, but a few species are found in unionids. Trichodina unionis is found in the mantle cavity of Anodonta cygnea and Unio spp. in Europe. Prevalence approaches 100% in some populations but with only about 10 per host. Diameter of T. unionis is about 70–100 μm. The most common location of this organism is on the labial palps, and less often the gills. Trichodina sp. was observed in unionids collected in Illinois and North Carolina. Histological examination did not reveal lesions associated with Trichodina sp. Other ciliates of unionids are the scyphidiid peritrich Mantoscyphidia sp., and low numbers of a scuticociliatid ciliate on the gills of Elliptio complanata in North Carolina.

Among Metazoan parasites, larvae of nematodes belong to the genus Hysterothylacium parasitizing the pericardial cavity of Diplodon suavidicus from the Amazon basin were reported by Lopes et al., but no lesions pathogen-specific are described.

On the other side, Trematodes of different families are reported in this group of animals as mussels result as intermediate hosts for digenean trematodes. The family of Aspidogastridae commonly parasitize freshwater mollusks. Four species of aspidogastrids have been reported: Aspidogaster conchicola, Cotylaspis insignis, Cotylogaster occidentalis, and Lophotaspis interiora. Two of these species, are among the most common symbionts of unionids, are widely distributed, and are found in several hosts in North America. Bucephalid trematodes in unionids belong to the genus Rhipidocotyle spp., recently called Bucephalus polymorphus. In Unio pictorum, Baturo found sporocysts of Rhipidocotyle campanula and provided a detailed description of the developmental stages of this parasite. In Europe two species of Rhipidocotyle in the unionid Anodonta anatina are reported: Rhipidocotyle campanula and Rhipidocotyle fennica. In North America the Rhipidocotyle spp. identified as parasite unionids are Rhipidocotyle septpapillata and Rhipidocotyle papillosa. The most serious effect of bucephalid trematodes is host sterility with gonadal tissues replaced by sporocysts also accompanied to follicle fibrosis. Additional lesions also can occur at kidney level.

Moreover, water mites like Unionicola spp. commonly occur as symbionts of freshwater mussels. More than half of the described species are considered as symbionts and in 2013 Edwards and Vidrine published a book on the topic, with information on biogeography, classification, mussel-mite interactions, coevolution and phylogenetics. Three genera are known as symbionts of freshwater molluscs, like Dockovdia, Najadicola, and Unionicola and generally they live on the gills, mantle or foot of their hosts.

Neoplastic Diseases

During the past 50 years a considerable literature has been published on spontaneous and experimentally-induced tumors in invertebrates. Invertebrate neoplasia have been described in different taxonomic groups, like sipunculids, annelids, ascidians, arthropods, insects and bivalve with economic interest. In particular, gonadal and haemic neoplasia of marine bivalves are the most common, and present the characteristics of malignant tumors. Others, less frequent type, are the tumors arise from epithelia, muscle and connective tissue, mostly classified as benign, with neither invasive behavior nor mitotic figures. These neoplasms like polypoid growths of the foot, mantle and pericardium have been found repeatedly in Unionid molluscs like Anodonta cygnea , A. implicata and A. californiensis. A fibroma, lined by simple columnar ciliated epithelium, arising from the palp of the mussel A. implicata, have been reported. Williams reported a pedunculated tumor, composed of glandular and muscle cells while Collinge also observed two tumors from the same species from the same species of freshwater mussel, with no microscopical descriptions.

Diseases and Control Measures

DISEASE	AGENT	TYPE	SYNDROME	MEASURES
Parasitic infection	Marteilia maurini	Protozoan	Potentially lethal; haemocyte infiltration of digestive gland (connective tissue and epithelia); extensive destruction of the digestive gland in heavy infections.	No curative measure; prevention & site selection; monitoring mussel transfer.
Viral disease	Picornaviridae-like virus	Virus	Heavy mortalities	
Vibriosis	Vibrios	Bacteria	Not specified	
Rickettsiosis	Rickettsia-like organisms; Chlamydia-like organisms	Bacteria	Microcolonies in the epithelial cells of the gills and digestive gland.	No curative measure; prevention & site selection.
Various parasitic infections	Steinhausia mytilovum	Micro-sporidian	Infects cytoplasm of mature mussel ova; incites a strong haemocyte infiltration response.	No curative measure; prevention & site selection; monitoring mussel transfer.
	Cliona	Sponge	Penetrates the periostracum forming holes in the outer surface and a tunnel network throughout the shell.	None
	Prosorhynchus sp.	Bucephalid trematode	Mantles show abnormal colouration (patchy yellow-white) in heavily infected individuals; castration; weakness; gaping.	None
	Polydora ciliata	Polychaete annelid	Burrows & blisters; mortalities; reduced condition index; loss of market quality.	None

Pea crab parasites	Pinnotheres pisum	Crustacean	Reduces market value	No curative measure; decrease stocking density.
Red 'worm' diseases	Mytilicola intestinalis; Mytilicola orientalis	Copepods	Usually commensal but may retard growth.	

With regard to shellfish regulations, preventive measures aim to limit imports only from countries where no outbreak of disease occurs according to the list specified by OIE International Aquatic Animal Health Code (notifiable pathogens). Although not a notifiable pathogen, the protistan parasite Marteilia maurini is hosted by the blue mussel, in contrast to the notifiable M. refringens, the oyster epizootic parasite. This indicates that mussel movements are not affected by legal regulations on notifiable diseases. Although high mortalities caused by parasites or infectious diseases have not yet been encountered in M. edulis, several parasites may be potentially harmful. Mussel transfers with parasites must be conducted with caution. Monitoring M. edulis populations and parasites is critical to prevent and limit associated risks.

Shrimp Culture

Many factors must be considered when a farmer is deciding which species of shrimp he should culture. Due to its large size and high price, Penaeus monodon and P. indicus are generally considered for farming. It has also been seen that both these species are suitable for farming in Kerala's environment. Apart from these candidate species other commercially important species such as Metapenaeus ensis, M. monoceros, M. brevicornis, Penaeus semisulcatus and P. merguiensis are also potential species that can be grown. Another potential candidate species that is flooding international market is the White leg shrimp, Penaeus vannamei. Many Asian countries have already started to culture this species.

Advantages of P. Monodon

- It attains a large size. Shrimp with a size of 10 to 12 pieces/kg are common, and sizes of 5 to 7 pieces/kg have been grown in ponds.
- It is the fastest growing of all shrimp tested for culture. In ponds, juveniles of 3 cm in length have been grown to a size of 75 to 100 g in only five to six months.
- Due to its large size, it brings a high price to the farmer. At peak seasons it demands over Rs. 450 per kg.
- It can tolerate a wide range of salinity, 0.2 to 70 ppt. Salinity within the range of 10 to 25 ppt has no appreciable effect on growth when food is sufficient. Growth is reported to be slower at very low salinities.
- It can tolerate temperatures up to at least 37.5 °C. Mortalities occur at temperatures below 12 °C only.
- It grows rapidly when fed either with animal or vegetable protein.

- Food conversion ratios are favourable. Values as low as 1.8:1 have been reported from Taiwan.
- It is hardy and not greatly disturbed by handling.

Tiger Prawn (*Penaeus monodon*)

Advantages of P. Indicus

- This shrimp grows to a fairly large size and brings a good price.
- It is fairly fast growing, especially when young. Cultured in tanks at a density of 15/m^2, it reached a size of 14 g in 16 weeks. In polyculture with milkfish in earthen ponds, females grew to about 28 g and males to about 12 g in 160 days.
- Survival is high during the first three months of growth or up to a size of about 10 cm.
- Wild seeds are usually abundant in estuaries near areas where the adults are present.
- Gravid females are relatively easy to obtain from the wild in numbers sufficient to operate a hatchery.
- Females can be matured in captivity with relative ease.
- This shrimp moves out of a pond with water discharge, making harvesting easy.
- Good growth has been obtained in intensive culture with a feed having 40 percent protein, which is lower than that required for some other species.
- The exoskeleton is relatively thin, giving greater portion of edible meat to total weight.

Penaeus indicus

Advantage of P. Vannamei

- Penaeus vannamei has the potential to grow as fast as P. monodon (at up to 3 g/wk) up to 20 g under intensive culture conditions.
- They are amenable to culture at very high stocking densities of up to 150/m2 in pond culture, and even as high as 400/m2 in controlled recirculated tank culture.
- Tolerates a wide range of salinities, from 0.5-45 ppt, is comfortable at 7-34 ppt, but grows particularly well at low salinities of around 10-15 ppt.
- P. vannamei is very tolerant to low temperatures (down to 15 °C) enabling them to be cultured in the cold season.
- P. vannamei require lower protein feed (20-35%) than P. monodon resulting in a reduction in operational costs and amenability for closed, heterotrophic systems and has a better Food Conversion Ratios (FCRs) of 1.2.
- Specific Pathogen Free (SPF) brood stocks are available for this species to produce disease free larvae.

P. vannamei

Site Selection

Selection for a suitable site is a critical activity and must be carefully determined before establishing of a shrimp farm. Site evaluation is not only undertaken to determine if a site is suitable for shrimp farming. It is also valuable in determining what modifications are needed concerning layout, engineering, and management practices to make shrimp farming possible at a given site. No site will have all the desirable characteristics, so a number of judgements have to be made for every site.

Topography and Climatic Condition

Topographically, the best areas for shrimp culture are those with average natural ground elevations of about 1-3 m above mean sea level or at least 1 m above the highest high tide level to allow drainage and harvesting. The cost of pumping increases in highly elevated sites. The sites should have minimum vegetative cover, be near the sea or other natural waterways such as rivers and streams, have easy access to roads, a sparse population and be nearly square or rectangular in shape.

In terms of climatic conditions, areas having short and not as pronounced a dry season with moderate rainfall distributed throughout the year are the best suited for shrimp farming. A pronounced long dry season may cause an increase in water temperature and salinity, which will promote excessive growth of algae and result in oxygen depletion at night.

Infrastructure

1. Accessibility: The farm must have good accessibility either by road or water, and communication systems throughout the year in order to facilitate supervision and transport of materials and products. It is important that the farm be within 3-6 hrs traveling time from the hatchery to avoid excessively long transportation time of the larvae and should be within 10 hrs from the processing plant to avoid deterioration of the product.

2. Electricity: Availability of relatively cheap and reliable power source is a major consideration in site selection. In areas where electricity supply exists, it is practical and beneficial to utilize electric power to run the farm, especially for the intensive culture system. It is advisable to have a back-up electricity generator as a secondary power source.

3. Security: Areas free from security risks result in favorable working conditions, productivity and less extra costs.

4. Availability of Labor and Other Factors: The availability of labor, equipment and commercial feed and supplies ensure smooth operations and successful crop.

Water Supply

Site should have good pollution free water supply of both freshwater and brackish water. Water quality parameters required for maximum feed efficiency and maximum growth of Penaeus monodon are given below:

Water Parameters	Optimum level
Dissolved Oxygen	3.5-4 ppm
Salinity	10-25 ppt
Water Temperature	26-32 (°C)
pH	6.8-8.7
Total nitrite nitrogen	1.0 ppm
Total ammonia (less than)	1.0 ppm
Biological Oxygen Demand (BOD)	10 ppm
Chemical Oxygen Demand (COD)	70 ppm

Transparency	35 cm
Carbon dioxide (less than)	10 ppm
Sulphide (less than)	0.003 ppm

Water from polluted areas containing high concentrations of suspended solids and organic wastes such as effluent water from industry; urban areas, agricultural and other farm locales should be avoided. A settling pond or a large reservoir should be used in such areas for sedimentation and treatment. In saline areas, a source of freshwater is useful for adjusting the salinity in the ponds and for domestic use by farm staff. The freshwater should be good enough for use and adequate throughout the year.

Soil Conditions

The type of soil is the most critical in site selection, since the shrimp will spend most of their time on the pond bottom during the culture period. Usually, clay or loam-based soil containing more than 90% clay and pH between 6.5-8.5 is preferable. Sites with sandy or silty soil should be avoided due to their porous nature that may lead to erosion, seepage of water and easy infiltration of waste into the soil. Prior to construction of ponds, samples of soil should be taken randomly from 5-10 spots at the surface and at 1 meter deep and sent to a laboratory for the analysis of soil texture and pH. Such data will be useful during pond construction and preparation.

Mangrove or acid sulfate soils are not suitable for shrimp pond culture due to their high organic matter contents and acidic nature that require a high water exchange rate and low stocking density. A pond built on mangrove soil will also encounter the problems of hydrogen sulfide and ammonia accumulation in the pond bottom. In the acid sulfate soil areas, the soil will develop high acidity when dried and then flooded which will lead to difficulty in stabilizing the pH of the pond water and in inducing the growth of plankton during the culture period.

Pond Design and Construction

A shrimp pond should be designed according to the characteristics of the selected site and the culture system. There is no unique design, but optimum and functional farm layout plan and design should be based on the physical and economic conditions prevailing in the locality.

Culture Systems

There are three types of shrimp culture being practiced in most countries:

Traditional/Extensive Cultures

The ponds have irregular shapes and sizes, mostly 1.5 ha and bigger with a peripheral ditch or canal of 4-10 m wide and 40-80 cm deep. The pond bottom may not be properly leveled, but tree stumps are usually removed, although this is not required. Ponds are normally filled with gravity flow water during the high tide period with natural seeds and left for 60-90 days, without additional seed stocking and feeding. Stocking density in this type of culture is 0.5-5.0 pcs/m^2. These ponds are normally partially harvested.

Semi-Intensive Culture

Ponds of 1-1.5 ha in size and are constructed with dikes to hold the water 1-1.5 m deep. PL are stocked into the pond at 10-15 PL/m^2 and fed with commercial diets and/or fresh diets. The shrimp are harvested at 90-120 days after stocking.

Intensive Culture

The ponds are usually of 0.5-1 ha in size and are designed to keep the water at 1.5-2.0 m deep. A reservoir of at least 30 % of the pond area is usually required. High stocking density of 25-60 PL/m2 with feeding rate of 4-6 times daily and strong aeration is maintained.

Open System

This system requires a high supply of good quality water because it needs a water, exchange of more than 20% of the total pond volume at one time, in order to reduce pond wastes and the density of the plankton. Seed can be stocked up to 60 PL/m^2 and will grow to 25-35 grams within 120 days. The open system has recently become less favorable to farmers since the environmental conditions, especially the quality of water, tend to deteriorate with time.

Re-circulation System

In order to avoid deterioration of the environmental conditions, several advanced and company run farms have adopted the re-circulation system to minimize contact with poor quality water from outside the farm. However, the farm must devote 40-50 % of the area for the construction of water storage/reservoir, sedimentation pond, treatment pond and drainage canals. To operate the system, cleaned seawater is initially pumped into the pond and kept within the system. During the culture period, the effluent from culture pond is drained into the sedimentation pond, treated with chemicals and pumped into the reservoir for re-supply to culture ponds. The stocking density for this system generally varies between 30-50 PL/m^2 and the culture period is between 110-130 days.

Minimal Water Exchange System

The majority of small farms cannot support space for construction of the water treatment pond and reservoir as in the case of the re-circulation system. To reduce contact with the water from outside the farm, the minimal water exchange system or closed pond system is practiced in some countries, particularly in Thailand. The system involves filling up the pond with cleaned seawater, treating it with chemicals to eradicate predators and competitors. Then the shrimps are stocked up to 30 PL/m^2 and cultured for a period of less than 100 days to attain the average weight of 10-20 gm. Since the system does not require water exchange, but maintains the water level in the pond by replacing the water loss due to evaporation and seepage with seawater or freshwater, it can be operated anywhere, even in the inland area where seawater is not easily accessible. The disadvantages of this system are that it requires low stocking density and high efficient water and waste management. However, it is suitable for production of small size shrimp because the culture period is limited.

Farm Design

An extensive shrimp farm should be of the size 0.4 - 0.5 ha and preferably drainable from the management point of view. The ponds generally should have concrete dikes, elevated concrete supply canal with separate drain gates and adequate life supporting devices like generators and aerators. The design, elevation and orientation of the water canals must be related to the elevation of the area with particular reference to the mean range of tidal fluctuation. The layout of the canals and dikes may be fitted as closely as technically possible to existing land slopes and undulation for minimizing the cost of construction.

Water Supply System

A shrimp pond is filled with water mostly by pumping. The pumps should be installed at locations where they can obtain water from the middle of the water column with least sedimentation and pollution. The pumps and inlet canal should be large enough to allow the ponds or the reservoir to be filled within 4-6 hrs. A screen should be installed at the inlet canal prior to the pumps to prevent clogging at the inlets.

Reservoir

A reservoir is important for the control of pond environment and storage of water supply when the water quality is inconsistent or the supply is intermittent. It is recommended that the area of a reservoir within a farm should be about 30% of the total farm area in order to hold a sufficient volume of the water supply. Some farms may use part of the reservoir for sedimentation purpose where biological filter feeding organisms are stocked. The reservoir must have an outlet that can allow total drainage.

Supply Canals

An intensive shrimp farm should have a water supply canal to convey the water from the reservoir to the ponds by gravity or pumping. The size of the supply canal will depend on the size of the culture pond, the efficiency of the pump and the required water exchange rate.

Ponds

A well-designed pond will facilitate the management of water exchange, harvesting of the product, waste collection and elimination, and feeding.

Shape

The shapes of pond that are found to be effective for shrimp culture are rectangular, square and circular. A well-designed pond is one that would allow circulation of the water such that wastes will be accumulated at the center of the pond. Some farmers improve the water movement in the square and rectangular ponds by making the corners of the pond rounded through addition of soil.

Size

Smaller ponds are easier to manage but the construction and operation can be costly. Ponds of 0.5-1.0 ha. are commonly used in intensive culture and 1-2 ha for semi-intensive culture.

Dikes

Earthen dikes, with or without lining, are found to be the most economical. Dikes should be designed to impound higher than 1 m depth of water and must be high enough to prevent flooding during the rainy seasons and the highest high tide. The slope of the dike depends on the nature of the soil. A slope of not less than 1:1.5 is normally used in the sandy soil area to avoid erosion and 1:1 is used for clay soils. One must be aware that shallow slopes will encourage the growth of benthic algae which will impair the quality of the water in the pond. Some dikes in a farm may be wider than the others to provide space for the access road, storage, electricity and aerators.

Pond Lining

Lining materials are used in pond where the soil contains a high percentage of sand, and organic matter and is acidic in nature. Lining can reduce erosion, water seepage, waste accumulation in the soil and the leaching of ammonia, hydrogen sulfide, acidic compounds, iron and other potentially stressful compounds into the ponds. The lining also allows easy removal of wastes from the feeding areas, reducing the time and costs to clean the ponds between cycles. Several lining materials are currently available. The economic life of liners varies according to the maintenance and the duration of exposure to sunlight.

Among the liners, laterite soil is less expensive and commonly used in shrimp farms. However, laterite soil liner may allow the penetration of wastes and requires effective cleaning up. Pond liners with PVC plastic sheeting and geotextiles can reduce the cost for aeration and cleaning up due to the easy movement of wastes and uneaten food on the smooth surface. The disadvantages of PVC plastic and geotextile-lined ponds are difficulties in maintaining plankton bloom within the first month of culture, problem of tears and the floating of the liner if the water and gas accumulate underneath them.

Gates for Inlet and Outlet

Each shrimp pond should have at least one gate for filling and draining water. However, a typical pond of 0.5-1 ha usually consists of two gates having similar structure for the inlet and outlet gates. The size of the gate is dependent on the size of the pond, but must allow the pond to be filled or drained within 4-6 hrs. Gates of 0.5-1.0 m wide are usually constructed, since gates wider than 1 m will cause difficulty in screening and will allow strong currents which will cause erosion of the soil. The position of the outlet should be at the lowest point of the pond with a gradual slope of 1:200 from the inlet to allow total drainage of the pond during harvesting.

The conventional gates constructed at the side of the pond should have a double screen, with fine a mesh for the initial period of culture and a coarser one for a later period. Some farmers may place both meshes in a single frame and cut out the finer mesh when the size of the shrimps are larger than the opening of the coarser mesh.

Central Drain

This has been employed in some farms and consists of perforated pipes laid horizontally at the center of the pond and connected to a pipe leading to the outlet. A screen of small mesh size is used to cover the drain for the first 50 days of culture and is removed to allow for easy removal of water when the shrimps are larger than the diameter of the pipe. This method has the advantages in that it can remove the waste and clean the pond bottom any time throughout the culture period.

Drainage Canal and Sedimentation Pond

The drainage canal of a shrimp pond should be at least 50 cm lower than the lowest point of the pond to allow drainage by gravity. The effluent will be drained into a sedimentation pond to settle the particulate wastes before water is pumped into the reservoir or released out of the farm. It is recommended that the sedimentation pond should be approximately 5-10% of the culture area and should be deep enough to prevent mixing and re-suspension of the wastes. Baffles or soft walls made of fine mesh net or plastic sheeting supported by stakes driven into the pond bottom, may be constructed in the sedimentation pond to decrease the velocity of water and increase the retention time which will enhance the settlement of the wastes. The wastes in the sedimentation pond should be removed periodically and discharged into the waste dumping area.

Waste Dumping Area

A shrimp farm should provide 5-10% of the area for dumping of the wastes. Wastes from the pond must be collected carefully and dumped into this area without discharging to nearby areas, which will contaminate the natural resources.

Buildings

Accommodation, storage, shop and guardhouses may be built in the farm as required. It is advised that accommodation for workers should be set up at various points around the farm for security purposes and to allow the ponds to be adequately monitored.

Pond Management

Pond Preparation

Before a pond can be stocked for a new crop, the excessive wastes, which accumulate in the pond during the previous crop, must be removed and the soil and water conditioned. Growing of shrimp in an improperly prepared pond may lead to difficulty in pond management during the culture period, which could result in a decrease in production capacity of the pond.

Pond Cleaning

The cleaning of a pond or removal of the wastes, especially the organic and phosphatic wastes that have accumulated in the pond bottom could be, accomplished by drying, liming and ploughing. However, these methods could still leave an adverse effect on the water and soil quality in the pond, which could result in a decrease in the production capacity of the pond.

There are two methods for cleaning a pond according to the possibility of the pond to be dried:

1. Dry Method: This method is used when the pond bottom can be dried completely. The pond is drained and left to dry in the sun for a period of 10-30 days. Then the waste is removed, either

manually or mechanically, and transported to the waste dumping area. Removal of waste by machines has an advantage that it can compact the bottom soil. However, this cleaning method by drying may lead to development of acidity, lowering the level of the pond bottom and the diffusion of wastes if the workers are inexperienced.

2. Wet Method: In areas where the pond cannot be dried completely, pressure washing can be used to flush out the wastes. This method takes a shorter time and is more efficient than the dry method. Flushing should be continued until the acid and dark anaerobic layer in the soil are removed. This method is suitable in the acid sulfate areas where the oxidation of the soil must be avoided. However, the method requires a sedimentation pond to all settlement of the suspended wastes to avoid contaminating the drainage canal and the natural environment. The remaining pathogens in the ponds can be eliminated during the liming process.

Liming

Once the pond is cleaned, it is then filled with water and left overnight before flushing out to remove debris and elevate the pH. This process should be repeated until the pH of the water remains above 7, and only then the lime is applied. The types of lime to be used depend on the water pH. It is recommended that agricultural lime ($CaCO_3$) or dolomite [$CaMg(CO_3)_2$] should be used in a pond with water pH near neutral and the hydrated lime [$Ca(OH)_2$] should be used in a pond with water pH below 5. The amount of lime to be used should be carefully calculated to avoid inducing an excessively high water pH, which may increase ammonia toxicity and result in the mortality of the shrimps.

The lime requirement of a pond depends on the soil pH. The measurement of the soil pH should be determined either by the wet soil method or by the dry soil method. During the application, lime should be spread throughout the pond bottom and up to the top of the dike. A large portion of lime should be applied over the feeding areas and to all parts of the pond that have remained wet. When the pond is properly limed and filled with water, the average water pH should be between 7.5-8.5 with daily fluctuation of less than 0.5. Agricultural lime, dolomite or hydrated lime at 100 kg/ha/day should be added to maintain the required pH.

Eradication of Predators

After liming, the pond should be filled to the maximum depth through a screen with fine mesh to prevent the predators and competitors from entering the pond. These animals, including fish, crustaceans and some invertebrates, may compete for food, prey on the shrimp or carry diseases and parasites. They may establish themselves in the pond that is not effectively screened effectively or is left for a long period of time.

Some chemicals should be used to eradicate these animals in the pond before stocking. Fish can be killed by the application of tea seed powder at the rate of 20-30 ppm. After the application of tea seed, the pond should be left for 3 days before the post larvae can be stocked. Tea seed may also be used when the shrimp has reached a weight of more than 2 gm. However, it must be remembered that tea seed is more toxic at high salinity and temperature, but less toxic at high pH. Application of tea seed in the evening may reduce pH and result in plankton die-off.

Snails can be eliminated by the application of quick lime (CaO) at 530 kg./ha and sun dried for 2-3 days. Then the pond should be thoroughly cleaned, filled with water and the other pests eradicated. Hypochlorite, either calcium or sodium salt, is currently used at 15-20% (60% active ingredient) to eliminate both vertebrates and invertebrates. The pond must be cleaned prior to the application of hypochlorite since hypochlorite may react with the organic matters and produces the toxic organochlorine compounds. Hypochlorite should be applied after the pond is filled to the maximum height and left for 3 days to allow the hatching of planktonic organisms. It should be remembered that hypochlorite should be used prior to the liming since the effectiveness of hypochlorite will be lowered in high pH conditions.

After the hypochlorite application, the pond should be aerated and the application of lime and fertilizer should be conducted on Day 3, while the PL can be stocked on Day 7. During the first month, water must not be added to the pond, unless the water quality is poor, to prevent the introduction of competitors and predators.

Fertilization

The pond must be fertilized with either organic or inorganic fertilizer to stimulate the plankton bloom in order to provide shade to the pond bottom and utilize the nitrogenous and phosphate wastes within the pond. The shade will also prevent the growth of harmful benthic algae. The sun dried chicken manure is the most common organic fertilizer to be used in the amount of 200-300 kg/ha. The manure must be soaked in water for 24 hours before it is spread over the surface of the water.

Inorganic fertilizers, such as urea (46% N) and compound fertilizers like, ammonium phosphate (16:20:0) or those with N:P:K combination of (16:16:16) can be used at 20-30 kg./ha. The fertilizer must be dissolved in water before it is spread over the water surface to avoid precipitation of the fertilizer onto the pond bottom, which will enrich the soil and accelerate the growth of benthic algae.

After fertilization, the plankton should bloom within a few days and the color of the water becomes slightly green. The fertilizer, either the organic or inorganic, should be applied daily in the pond at 5-10 % of the initial amount to maintain the plankton bloom. If the plankton has not bloomed within a few days, additional fertilizer must not be applied, but plankton rich water or green water from the reservoir should be added.

Aeration

A 0.5-1.0 ha pond would require, four aerators installed at the corners of the pond, approximately 3-5 m from the bottom of the dike and positioned at an angle that will encourage the maximum water flow within the pond. The type of aerator to be used depends on the depth of the water. One horsepower paddle wheel aerators should be used in ponds of less than 1.2 m water depth and the 2 HP (horsepower) paddle wheel aerators should be used in ponds deeper than 1.2 m. The most popular type of aerator is the long arm paddle wheel aerator, which is driven by a 2-10 HP electric, or diesel motor mounted on the dike. The aerators should be switched-on 24 hrs before the PL are stocked to allow enough time to create the current and clean up the feeding area.

Stocking

The most suitable species for culture are the white prawn Penaeus indicus and tiger prawn P. monodon. The stocking density varies with the type of system adopted and the species selected for the culture. As per the directives of Supreme Court only traditional and improved traditional shrimp farming can be undertaken within the Coastal Regulation Zone (CRZ) with a production range of 1 to 1.5 t/ha/crop with stocking density of 40,000 to 60,000/ha/crop. Outside CRZ extensive shrimp farming with a production range of 2.5 to 3 t/ha/crop with stocking density of 1,00,000/ha/crop may be allowed.

Seed Selection

Selection of good quality seed for stocking into a pond is the first important step of the shrimp grow-out management. The farmer must ensure that he or she gets healthy seed by purchasing them from reliable hatchery or hatcheries. It may not always be possible to obtain the desired shrimp seed due to limitations in availability and quantity.

The following parameters should be taken into consideration in purchasing shrimp seed for stocking:

1. Size: Seeds of PL 15-20, indicated by the appearance of 4-6 spines on the rostrum, are recommended for stocking in a pond. The healthy PL should have the muscle-to-gut ratio in the sixth abdominal segment of about 4:1 or the thickness of the gut should be about the thickness of the muscle. Practically, seed from the first and second spawning of a broodstock with uniform size can be used.

2. Morphology: The post larvae should have normal appearance of trunk, appendages and rostrum. The abdominal muscle must be clear, no discoloration or erosion on any parts of the body, the gut should be full of food, and the muscle should fill the carapace.

3. Color: Post larvae with the presence of pigment cells in the uropods should be used since this indicates the stage of development. PL that will have high survival and growth rates will be light gray, brown to dark brown and black in colour. Signs of red or pink coloration are normally related to stress.

4. Behavior: Healthy seed swim straight, respond rapidly to external stimuli such as a tap on the side of the basin, actively swim against the current when the water is stirred, and cling to the sides rather than aggregate or be swept down into the center of the container when the current has subsided.

5. External Fouling: Seeds should be free from external parasites, bacteria and other fouling organisms. The presence of these organisms indicates unhealthy conditions, which will affect growth and survival of the PL. It is recommended that before purchasing, the farmer should visit the hatchery to check the seed once or twice either in the early morning or late afternoon, especially one day prior to stocking. However, healthy seed with some fouling may be used when the animals are in good condition after treatment.

6. Pathogen Free: Seed should be checked for the presence of viral occlusion bodies. Seed with large numbers of occlusions indicate stress conditions and will not so vigorous in the pond.

Stocking Density

When a farm is ready for operation, the optimum stocking density of PL in a pond should be determined in accordance with the production capacity of the farm and the culture system, which include the soil and water quality, food availability, seasonal variations, target production, and farmer's experience. It is recommended that farmers should start a new crop with a low stocking density to access the production capacity of the pond. If production is successful, then the stocking density could be increased for subsequent crops. Overstocking should be avoided since it may result in management problems and loss of entire production.

The stocking density between 10-20 PL/m^2 is usually practiced in a semi-intensive culture. In an intensive culture, a well-managed pond with consistent good water quality can stock up to 25-30 PL/m^2 at 1.2 m water depth and up to 40-50 PL/m^2 at 1.5 m water depth or deeper. However, it must be emphasized that intensive cultures involve high densities and can only be sustained in well-managed farms under an experienced farmer.

Technique of Stocking

Proper stocking techniques will prevent unnecessary mortality of seed. The following methods have shown excellent results.

1. Transportation: Seed are normally transported in plastic bags. The bags are usually filled up to 1/3 with water, oxygenated and then placed inside styrofoam boxes. If the transportation is longer than 6 hours, small bags of ice should be added into the boxes to reduce the water temperature and maintain it at 20-22 °C. The densities of PL in a bag are 1,000-2,000 seed/l for PL 15 and 500-1,000 PL/l for PL 20. The ideal time for transportation is in the early morning or evening to avoid excessively high temperatures during the day, unless a covered vehicle is used.

2. Acclimation: To eliminate stress, the seed should be maintained in water of constant salinity for at least 1-week prior to transfer. The adjustment of salinity by about 3 ppt daily is advisable. Acclimation of seed to the water pH and temperature of the pond must be rendered upon arrival. Two common techniques are used for gradual acclimation of seed to the water conditions in the pond. The first method is accomplished by placing the seed and water from the transported bag into a tank at the side of a pond containing an equal volume of well-aerated pond water. The seed will be kept for 0.5-1 hr before being siphoned into the pond. The second method, the most favorable one, is to float the plastic bag in the pond until it has reached equilibrium. The bags are opened one by one and pond water is added gradually to an equal volume. After a further 30 min of acclimatization, the seed are released directly into the pond by distributing them throughout the area of the pond or into a nursing system. The actual numbers of seed at stocking can be estimated by counting the PL individually in 3-5 bags with a spoon or small net to attain the average number in each bag and multiplied by the total number of bags.

3. Nursing of Shrimp Post Larvae: To ensure high survival and adequate feeding of seed during the first 2-3 weeks, some farms may stock the PL in a separate nursing pond or a small impoundment, usually 5-10 % of the total pond area, within the culture pond. The nursing system will help in concentrating the seed in a limited area until they reach PL 30-40 and in more accurate monitoring for survival and feeding of the PL. However, it appears that the separate nursing pond system may lead to some unfavorable results in that the size of the PL varies widely, ('broken sizes'), and the seed difficult to harvest and would experience stress during harvest and transport to the culture pond. As a result, a farmer prefers to nurse the seed in an impoundment installed inside the pond, rather than in a separate pond. Recently, some farmers employ a system in which high densities of seed (100-200 PL/m^2) are stocked into a pond for 1-2 months, then approximately half of the juveniles are transferred to another pond by large lift nets. The same acclimation process should be performed during seed and juvenile stocking.

In a very intensive pond (stocking density greater than 30 PL/m^2) where the nursing impoundment is not available, a survival pen may be installed to estimate the survival of the seed during the first 2 weeks after stocking to allow accurate feeding management. The survival pen may be a small net pen or happa of approximately 1 m^2 containing 100 seed or a large net pen of usually 10 m2 at 100 PL/m^2 stocking density. In the small pen, the seed can be counted accurately while the seed in the large pen may be counted by using a 1 m^2 lift net placed with 10% of the feed. In this method, seed should appear in the lift net at 3-4 days after stocking and the number of shrimp in the net should be counted at 2 hours after feeding once daily. The survival number of shrimp can then be estimated.

If the survival rate during the nursing period is less than 50%, the problems that cause this initial mortality must be identified and rectified and the addition of more seed should be considered. Seed can be added up to 30 days post-stocking without causing a variation in size at harvest. If the survival is less than 30%, the pond should be drained and prepared for a new crop. Some farmers release seed directly into the pond. In this direct stocking method, the survival number of seed during the first 2 weeks post stocking may not be accurately estimated, since the shrimp will not approach the feeding trays during this period.

Feed and Feeding

Cost of feed constitutes a major part of the production cost and accounts for 50% to 70% of the total variable cost. The use of feeds will improve shrimp production and increase profits. The availability of nutrients from feeds depends on the type and quality of the raw material used, the formulation, the feed processing, feed storage conditions and the feeding management. Therefore, feed and feeding practices for semi-intensive or intensive shrimp farming require a basic understanding of nutrition and feed requirements.

Shrimp diets may be supplementary or complete. In a extensive system the shrimps need a complete diet. Although natural food items have good conversion values but they are difficult to procure in large quantities and maintain a continuous supply. At present most of the aquaculture farms depend on imported feed with a FCR (Food Conversion Ratio) of 1:1.5 - 1.8. The feeding could be done by using automatic feed dispensers, or by broadcasting all over the pond. If feeding trays are employed in selected pockets in the pond wastage in feed can be reduced.

Feed evaluation using check trays

Nutrient Requirements

Shrimp requires approximately 40 essential nutrients. These nutrients are provided in various amounts by natural food and supplemental feeds. Although the nutrition principles are similar for all animals, the quality and quantity of nutrient requirements vary from species to species. The recommended levels of nutrients and dietary components for black tiger shrimp are listed in table below.

Recommended nutrient levels for shrimp feed on percentage fed basis.

Shrimp size (gm)	Protein (%)	Fat (%)	Fiber (%)	Ash (%)	Moisture (%)	Calcium (%)	Phosphorus (%)
0.0-0.5	45	7.5	Max.4	Max.15	Max.12	Max.2.3	Min. 1.5
0.5-3.0	40	6.7	Max.4	Max.15	Max.12	Max.2.3	Min. 1.5
3.0-15.0	38	6.3	Max.4	Max.15	Max.12	Max.2.3	Min. 1.5
15.0-40.0	36	6.0	Max.4	Max.15	Max.12	Max.2.3	Min. 1.5

Protein

Shrimp consume dietary protein to obtain a continuous supply of amino acids for normal growth. About 20 major amino acids make up most of the proteins and 10 are essential including methionine, arginine, threonine, tryptophan, histidine, isoleucine, leucine, lysine, valine and phenylalanine. Thus, essential amino acids must be provided in adequate quantities and qualities (well-balanced) in the diet. On the other hand, the recommended dietary protein levels for shrimp vary from 30% to 55% depending on the shrimp size and species. It is believed that post-larval shrimp require a higher protein level than larger shrimp.

Lipid

The lipid requirement of shrimp depends on their essential fatty acids and phospholipid content. There are four fatty acids, which are considered essential for shrimp, namely linoleic (18:2n6), linolenic (18:3n3), eicosapentaenoic (20:5n3) and decosahexaenoic (22:6n3). In general, plant oils are high in 18.2n6 and 18.3n3, while the marine animal oils are high in 20:5n3 and 22:6n3. The phospholipid requirement is 2%; however if lecithin is used this level can be reduced to 1%. The requirement for cholesterol ranges from 0.25% to 0.4%. In addition, the recommended lipid level ranges from 6.0% to 7.5% and the level should not exceed 10 %.

Carbohydrates

The utilization and metabolism of carbohydrates by shrimp are limited. Their type and level in the diet have been shown to affect shrimp growth. Starch as the carbohydrate source is utilized better than dextrin or glucose for Penaeus monodon.

Vitamins

Little is known about vitamin requirements in shrimp. In intensive farming, vitamins must be supplied in the diet for normal growth. Commercial shrimp feeds are usually over-fortified with vitamins to overcome shortfalls due to processing loss and feed storage. The minimum requirement for vitamin C, which is easily lost, is about 50-150 ppm for Penaeus monodon.

Minerals

Shrimp can absorb or excrete minerals directly from the aquatic environment via gills and body surfaces. The dietary requirement for minerals is largely dependent on the mineral concentration of the environment in which the shrimp are being cultured. Among the other minerals, phosphorus is the most important, and is recommended at 0.9% as available phosphorus in the diet. Calcium is not considered to be a dietary essential. However, its level in feed needs to be monitored because it is important to maintain calcium to phosphorus ratio of 1:1 to 1:1.5. Calcium should not exceed 2.3% in the diet.

Feed Quality

The use of good quality feed will improve shrimp production and profits, and minimize the environmental pollution from shrimp farming. The practical indicators of good quality feed are:

1. Feed Conversion Ratio (FCR): An ideal FCR always results in model growth rate, healthy shrimp

and clean pond bottom conditions. Only the superior quality of feed can achieve an FCR of 1.2. According to recent data, an FCR as low as 1.2 has been achieved, but many farmers are still obtaining FCRs of higher than 2.2. Therefore, besides the feeding management, the FCR is also closely related to the quality of feed.

2. Attractability: The model quality shrimp feed must be highly palatable.

3. Good Water Stability: Since shrimp are a slow feeder animal, the water stability of suitable feed should be over 2 hours for Penaeus monodon.

4. Packaging and Optimal Conditions Storage: Feed quality will rapidly deteriorate if feed is not packed well and properly stored. Feed should be stored in a dry, cool and well-ventilated place to maintain consistent moisture and temperature. Feed should not be stored in direct sunlight and should not be kept longer than 3 months from the time of processing. The spoiled or old feed should not be used.

Feeding Management

A high FCR or high amount of feed required to produce unit weight gain indicates overfeeding, and consequently, a poor FCR is usually associated with poor growth rate, low weight gain, stressed shrimp, mediocre water quality and adverse pond bottom conditions. Therefore, the proper amount of feed is the most critical factor of feeding management. The guidelines for feeding adjustment to be made according to the mean body weight of the shrimp are shown in table below. Since many factors are involved in shrimp feed consumption, careful and frequent observation of shrimp is the most reliable approach for determining the optimal feeding amount. There are many major key factors for successful intensive shrimp culture. Use of good quality feed with better feeding management by low feed conversion ratios and improved farm management are the important goals to farmers, not only for gaining greater profit, but also for minimizing the pollution of shrimp farming area.

Table: Recommended feeding rate for shrimp based on body weight.

Shrimp Live Body Weight (g)	Recommended Feeding Rate (% body weight/day)
2 – 3	8.0 - 7.0
3 – 5	7.0 - 5.5
5 – 10	5.5 - 4.5
10 – 15	4.5 - 3.8
15 – 20	3.8 - 3.2
20 – 25	3.2 - 2.9
25 – 30	2.9 - 2.5
30 – 35	2.5 - 2.3
35 – 40	2.3 - 2.1

Water Quality Management

Water conditions in the rearing pond deteriorate through the production cycle due to uneaten food, animal's excretion, etc. Generally, shrimp farmers use dissolved oxygen (DO), pH, ammonia, water color and water odor as indicators to judge the water quality of the pond.

These parameters are observed regularly by using either scientific equipment or the farmer's experience in order to control them within the optimum range.

Optimum Water Quality for P. Monodon Pond

Parameter	Value
Dissolved Oxygen	4 (mg/l)
pH	7.5 - 8.5
Ammonia	<0.1 (mg/l)
Transparency	30 - 45 (cm)

Dissolved Oxygen

The amount of oxygen dissolved in the pond water is vital to the shrimp's health. However, in the rearing pond, dissolved oxygen is mainly consumed by pond sediment (50-70%) and plankton (20-45%). Only a small portion of dissolved oxygen is consumed by the shrimp (5%).

The level of dissolved oxygen can be controlled in 3 ways. Firstly, by increasing the water surface area by means of placing paddle wheels in the right position. This is not only causes proper water circulation, but also adds oxygen to the pond water. Secondly, by controlling plankton density to an optimum level and thirdly, by minimizing excess organic substances, such as uneaten food.

pH Adjustment

Shrimp farmers control water pH within the optimum range of 7.5-8.5, and limit diurnal pH fluctuation to less than 0.5 by applying lime. The application of lime is as follows:

- At the beginning of a crop cycle, when water pH ranges between 7.5-7.8 about 4.8-8.0 kg/ha of dolomite should be used every 2-3 days.
- When pH is in the range of 7.5-7.8 and there is less than 0.5 unit difference between the pH in the morning and the pH in the afternoon, 4.8-8.0 kg/ha of dolomite should be used every 2-3 days.
- If the pH in the morning is less than 7.5, 4.8 kg/ha dolomite should be used every day until the morning pH is increased to above 7.5.
- If the pH in the morning is higher than 8.0 and the pH in the afternoon is higher than 9.0, 4.8-8.0 kg/ha of dolomite should be used every day until the day's pH difference is less than 0.5.

- In the second half of a crop cycle, 8.0 kg/ha of dolomite should be used every day or at least every 2 days, depending on water color.
- Every time before exchanging water, 4.8-8.0 kg/ha of dolomite should be used.

Water Color Control and Adjustment

The color of pond water mainly results from suspended particles of phytoplankton. Plankton density and species are two management aspects that require attention of shrimp farmers.

In the first 2 months of shrimp culture, an additional fertilizer either organic (10-30 kg/ha) or inorganic (1-3 kg/ha) is added to the pond in order to ensure that there are enough nutrients for plankton bloom. After this period, nutrients derived from uneaten food normally are at adequate levels. Too many nutrients in some cases may lead to excessive plankton bloom, followed by plankton crash. In an open shrimp culture system the farmers exchange pond water with natural clean seawater to reduce excess plankton density. But in a closed system, where exchanging water is not needed, shrimp farmers use algaecides such as calcium hypochloride or benzalkonium chloride (BKC) 0.1-1 ppm to reduce plankton density.

In cases where undesirable water color appears like the 'red tide' caused by certain types of plankton such as dinoflagellates, the plankton can be controlled by switching-off the aerators for a period of time and applying BKC (0.1-1 ppm).

Water Exchange

Mass shrimp mortality in a pond associated with deteriorating environmental conditions has occurred frequently during the last 5 years. Shrimp farmers have tried to solve this problem by changing the culture system to a low water exchange system, including partial water re-circulation, full water re-circulation and a closed system.

Partial water re-circulation shrimp farming system is practiced where a supply of good quality water may only be available for short periods of time. Normally, the farm area is divided into 4 portions: culture area (60-70%), effluent settlement (10%), mixing reservoir (5-10%), and inlet reservoir (15-20%).

In the full water re-circulation shrimp farming system, where seawater can be treated and re-circulated, the farm area is divided into the culture area (40-50%), inlet water treatment (15%), seawater storage reservoir (20-25%) and effluent settlement pond (15-20%).

In the close shrimp culture system or zero-water discharge system, no pond water exchange is needed. However, the aeration in the pond must be adequate for shrimp respiration and oxidation of organic waste. Additional seawater may be required to make up for losses in the system. The technique provides disease-free seawater with no effluent being discharged. Shrimp may grow slowly and furnish lower production than those of an open or water circulation system.

Harvesting and Handling

Successful harvesting can be achieved if the shrimp can be harvested in good condition within a short period of time. The harvesting technique should not damage or excessively contaminate the shrimp with waste. Rapid harvesting will reduce the risk of bacterial contamination and the shrimp will still be fresh when reaching the processor.

Complete harvesting can be carried out by draining the pond water through a bag net and hand picking. The average culture period required is around 120-150 days during which time the prawns will grow to 20-30 gm size (depending on the species). It is possible to get two crops in a year. Harvested shrimps can be kept between layers of crushed ice before transporting the consignment to market.

Methods of Harvesting

Two methods of harvesting are generally practiced on farms. These are either by draining the pond and catching the shrimp in a bag net or by netting the shrimp within the pond.

For the first method of harvesting, ponds and outlets should be appropriately designed and be able to completely drain the pond within 4-6 hrs. A bag net should be able to be fixed to the outlet to collect the shrimp that are carried by the out flouring water. The best time for harvesting is early in the morning and it should be completed before mid-morning. In ponds that can only be drained at low tide, the harvest should be conducted whenever possible. The shrimp should be regularly removed from the harvesting bag in small quantities to prevent damage.

When netting the shrimp within the pond, either a small electric net or a large seine net can be used. The water level of the pond should be reduced to 0.5-0.75 m deep and workers will need to go inside the pond for netting. This method is less advantageous the pond bottom will be disturbed, thus causing contamination of the shrimp. It is also slower and may take a long time to complete.

With either method, it is necessary to hand-pick the remaining shrimp in the pond, after the pond is drained. The harvested shrimp can be quickly killed by giving them a temperature shock (dip in iced water) to prevent damage and to improve storage.

Cast netting of Shrimps

Timing of Harvesting and Selling

The timing of harvesting depends on the condition of the shrimp in the pond and also the market price. Shrimps are sampled by a cast net from different areas of the pond to determine their average body weight and general condition. The proportion of soft shell shrimp should not be more than 5% at the time of harvest. This proportion could be obtained by scheduling the harvest halfway between two moultings. The time of moulting is indicated by the presence of exuviae in the pond. For example if the average body weight of the shrimp is 30 gm, then the harvest should be planned for 7-8 days after the exuviae are observed, as the next moulting cycle can be observed after 14-16 days. Harvested shrimp should be iced and transported to cold storage or processing plants in less than 10 hrs.

Quality Control

Before harvesting and/or exporting, shrimp should be examined for their health, hygienic quality and safety for consumers. Unhealthy shrimps, which are easily recognized through their appearance, will not be acceptable to consumers and market value could be reduced. Unhealthy shrimp should be treated before harvesting or removed during harvesting and processing if the proportion of unhealthy shrimp in the stock is low.

Human pathogenic organisms could contaminate the shrimp during harvesting, storage and processing. Therefore, samples of shrimp should be sent to a reliable laboratory to conduct necessary test to certify the hygienic quality of the products, before exporting or sending them to market. The harvested shrimp should also be checked for antibiotics and heavy metal residues before export. If the shrimp have been treated for unhealthy conditions with antibiotics, the recommended withdrawal period should be followed.

Banned Antibiotics, Pesticides and Pharmacologically Active Substances

1. Chloramphenicol	12.Endosulphan
2. Nitrofurans including Furazolidone, Nitrofurazone, Furaltadone, Nitrofurantoin, Furylfuramide, Nifuratel, Nifuroxine, Nifurprazine and all their derivatives	13. Sulfonamide (except approved sulfabromomethazine, sulfadimethoxine and sulfaethoxypyridazine)
3. Neomycin	14. Ronidazole
4. Nalidixic Acid	15. Ipronidazole
5. Sulphamethoxazole	16. Other nitroimidazoles
6. Aristolochia spp. and preparations thereof	17. Diethylstilbestrol (DES)
7. Chloroform	18. Dimetridazole
8. Chlorpromazine	19. Clenbuterol
9. Colchicine	20. Metronidazole
10. Dapsone	21. Fluoroquinolones
11.Nuvan	22. Glycopeptides

Disease Management in Shrimp

Cultured shrimps suffer from various diseases due to infectious and non-infectious causes. Infectious diseases are caused by viruses, bacteria, fungi and certain parasites. Treatment cannot be carried out effectively when shrimp diseases occur in a pond. The best way to get rid of diseases is by practicing good farm management or prevention. In this regard, information on various kinds of diseases and their prevention procedures are useful.

A. Virus Infection

Monodon Baculovirus Disease (MBV)

- Etiological Agent: MBV-type or PmSNPV is a type A occluded monodon baculovirus.
- Clinical Signs: Lethargy, anorexia, poor feeding, dark colouration and reduced growth rate. Infected shrimps are often associated with fouling of gills and appendages by ciliates such as Zoothamnium spp. and Vorticella spp. Acute infection leads to loss of epithelial cells of hepatopancreas.
- Treatment: No treatment available for MBV infection.
- Prevention and Control: There is little information on prevention and control of the MBV infection in shrimp pond culture. The prevention method for the MBV infection is possibly through avoidance by screening the PL's before stocking shrimp in the pond.

Hepatopancreatic Parvo-like Virus (HPV) Disease

- Etiological Agent: HPV is caused by a small parvo-like virus, 22-24 nm in diameter.
- Clinical Signs: Reduced feeding, poor growth rate, body surface and gill fouling with ciliates and occassional opacity of abdominal muscles. Severe infections may include a whitish and atrophied hepatopancrease, anorexia and reduced preening activity. Losses may be occur due to the increased occurance of surface and gill fouling organisms and secondary infections by the opportunistic Vibrio spp.
- Treatment: No treatment available for HPV infection.
- Prevention and Control: No information is available on the prevention and control procedures for HPV infection. However, screening the PLs before stocking shrimp by routine histology or the Giemsa-impression smear method is recommended.

Yellow-head Disease (YHD)

- Etiological Agent: Yellow-headed virus (YHV) is a ssRNA, rod shaped, enveloped virus with two rounded ends.
- Clinical Signs: The affected shrimp shows a marked reduction in food consumption. Following this, a few moribund shrimp will appear swimming slowly near the surface of the pond dike and remain motionless. The animals have pale bodies, a swollen cephalothorax with a light yellow to yellowish hepatopancreas and gills. A high mortality rate may reach 100% of affected populations within 3-5 days from the onset of disease.
- Treatment: No treatment is available for YHV infection.

- Prevention and Control: The reliable method to prevent the occurrence of YHD is possibly through avoidance, such as careful selection of post larvae, reduction or elimination of horizontal transmission including carriers, disinfection of contaminated ponds or equipment with 30 ppm; and chlorine, providing shrimp with good waterquality and proper nutrition.

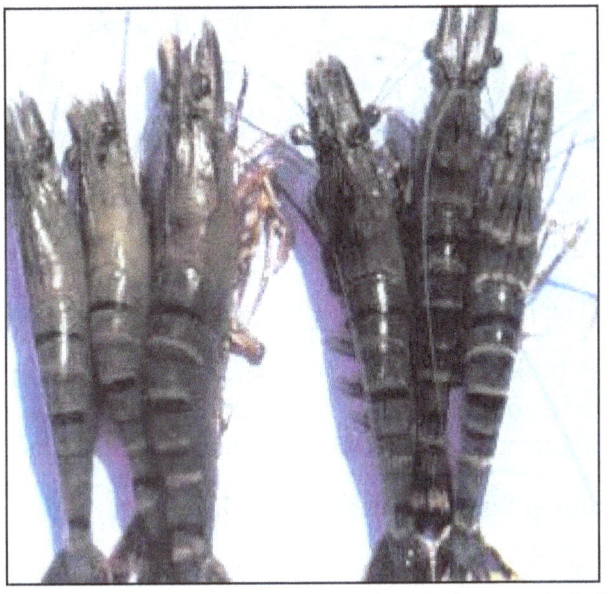

Yellow head disease (YHD) seen in three shrimps on the left.

White Spot Disease (WSD)

- Etiological Agent: The disease is caused by the dsDNA virus, Systemic Ectodermal and Mesodernal Baculovirus (SEMBV).
- Clinical Signs: Clinically affected shrimp were first seen to swim to the water surface and congregate at the pond dikes. Typical clinical signs include white spots or patches, 1-2 mm in diameter, on the inside of the shell and carapace, accompanied by reddish discoloration of the body. SEMBV is able to cause acute epizootics of 5-10 days duration with mortality rate from 40% to 100%.
- Diagnosis Procedure: The diagnosis procedure of SEMBV infection is based on the appearance of the intranuclear hypertrophy in stained histological sections and the presence of virus particles in the nucleus of the infected cells observed under the electron microscope. PCR technique is recently used to detect SEMBV in shrimp larval and other stages, including broodstock and subclinical virus carriers.
- Treatment: No treatment is available for SEMBV infection.
- Prevention and Control: Prevention practices through avoidance are strongly recommended for the farmers, involving the combinations of efficient pond management, use of proper feed, selection of good quality of PL, reduction of possible carriers, avoidance of introduction of contaminated water into the pond, and disinfection of all equipment and utensils.

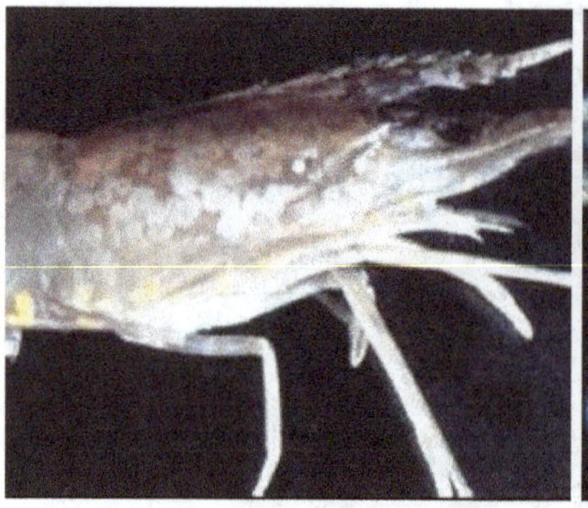
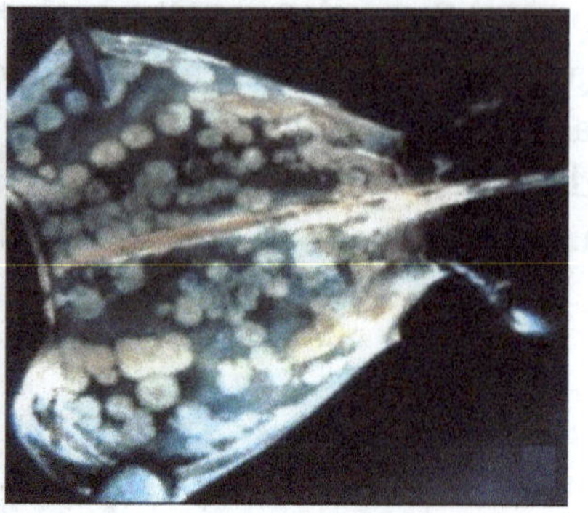

White spot disease in shrimp. White spots on the cephalothorax.

Infectious Hepatopancreatic and Lymphoid Organ Necrosis (IHLN)

- Etiological Agent: The primary cause of the disease is attributed to viral etiology.
- Clinical Signs: Light pinkish to yellowish discolouration of the cephalothorax region. Often fouling by ciliate protozoan Zoothamnium seen. Blackened and necrotic hepatopancreas. Secondary bacterial infection from bacteria such as Vibrio alginolyticus seen.
- Treatment: No treatment is available for IHLN infection.
- Prevention and Control: Keep the physico-chemical condition of pond environment within acceptable levels. To avoid bacterial and viral pathogen entering from outside, closed culture could be useful in prevention of IHLN disease.

B. Bacterial Infection

Luminous Vibriosis

- Etiological Agent: Vibrio harveyi, Vibrio vulnificus.
- Clinical Signs: High mortality rate in young juvenile shrimp (one month syndrome). Moribund shrimp hypoxic often come to the pond surface and edges of pond. Vertical swimming behavior immediately before onset of acute mortality. Presence of luminescent shrimp in ponds.
- Treatment: Disinfection of intake water with Formalin (100-200 ppm). Administration of Oxolinic acid (0.6 ppm) and Sarafloxacin (5mg/kg) through feed for 5 days.
- Prevention and Control: Proper pond and water management. Utilization of reservoir for intake water.

Vibriosis

- Etiological Agent: Vibrio vulnificus, V. parahemolyticus, V. alginolyticus, V. anguillarum, V. damsella, V. fluvialis and V. mimicus.
- Clinical Signs: High mortality rates, particularly in young juvenile shrimp. Moribund shrimp with corkscrew swimming behavior appear at edge of pond. Reddish discoloration of juvenile shrimp.

- External Fouling: Black spots, chronic soft shelling.
- Treatment: Disinfection of intake water i.e. formalin 100-200 ppm. Anti-microbial preparation application through feeds (Oxolinic acid 0.6 ppm and Sarafloxacin 5 mg/kg).
- Prevention and Control: Proper pond and water management and utilization of reservoir for intake water.

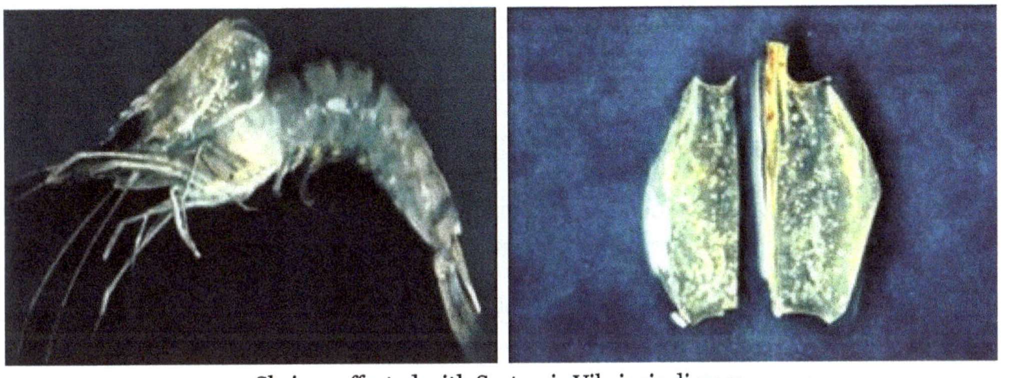

Shrimp affected with Systemic Vibriosis disease.

C. Fungal Infestation

Larval Mycosis

- Etiological Agent: Filamentous fungi of genus Lagenidium spp. and other filamentous fungi, such as Sirolpidium spp. and Haliphthoros spp.
- Clinical Signs: Eggs and larve are weak and appear whitish. Moratlities may reach 100% within two days. Fungal mycelium replaces the larval tissues and ramifies into all parts of the body and protrudes out of the body and develops into sporangia.
- Prevention and Control: General hatchery management practices such as use of UV sterilised and filtered seawater, adequate water exchange etc., must be strictly followed. Rearing water, equipment used in the hatchery and all hatchery facilities must be thoroughly disinfected before retarting the hatchery operations.

D. Protozoan and Parasitic Infestation

Black Gill Disease

- Etiological Agent: Fusarium spp.
- Clinical Signs: Brownish to blackish discoloration on the gills of juvenile shrimp.
- Treatment: No treatment is available for fungal infestation without harming the shrimp.
- Prevention and Control: No information on prevention and control. However, good management of the pond bottom and prevention of the entry of wild crustaceans into the pond, which may carry pathogen, can be effective control practices.

Black gill disease.

urface Fouling iseases

- Etiological Agent: Many species of bacteria, algae and protozoa such as filamentous bacteria, Leucouthrix sp., Flavobacterium sp. and Zoothamnium sp.
- Clinical Signs: Infected shrimps show black/brown gills or appendage discoloration or fuzzy/cottony appearance due to a heavy colony of the organisms. In some cases, the severely affected shrimp die during the molting period.
- Treatment: Chlorine and formalin are often used to treat those commensal organisms if shrimp display heavy infection. Changing water is the most preferable management, which stimulates molting of the shrimp in order to reduce the infestation.
- Prevention and Control: Prevention and control of the occurrence of surface fouling are usually done through maintenance of good sanitary conditions at the pond bottom and the overall pond area. Organic matters and suspended solids in the pond should be reduced to prevent the attachment of those fouling organisms. This is achieved by changing the water or applying lime.

Microsporidosis (Cotton Shrimp Disease or Milk Shrimp Disease)

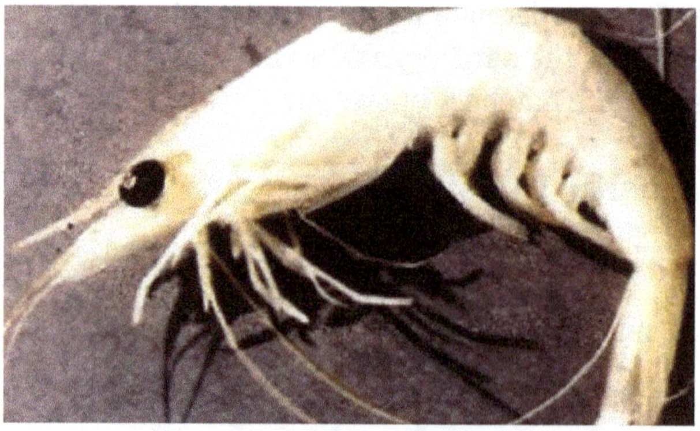

Cotton shrimp disease.

- Etiological Agent: Microsporidia such as Thelohania spp., Nosema spp., and Pleistophora spp.
- Clinical Signs: Infected shrimps appear opaque and cooked. Gradual and low levels of mortalities are observed. Microsporidia invade and replace gill, muscle, heart, gonads and hepatopancreas, and cause necrosis in these regions.

- Prevention and Control: Maintain of good sanitary conditions at the pond bottom and the overall pond area.

E. Non-Infectious Diseases

Soft-shell syndrome

Shrimps with persistent soft shell disease.

- Etiological Agent: The exact cause of soft-shell syndrome is not known. However, low saline condition in the culture pond and deterioration of pond bottom condition are some physico-chemical factors that causes this disease. Shrimps fed with low protein diet, contamination through agricultural run-off, high soil pH, low water phosphate and low organic matter in soil all have an impact on soft-shell disease.
- Clinical Signs: Shrimps are weak, usually off-feed, have a loose thin exoskeleton. Rostrum is stiff as healthy shrimps. Wavy undulating intestine is clearly visible.
- Prevention and Control: Low stocking density, feeding with high quality feed and frequent water exchange are likely to reduce the recurrence of the disease.

References

- Disease-Agents-and-Parasites-of-Carp: researchgate.net, Retrieved 9 March, 2019
- Fish-seafishes, fishery: tnau.ac.in, Retrieved 19 June, 2019
- Oncorhynchus, fishery: fao.org, Retrieved 30 April, 2019
- Mytilus-edulis, fishery: fao.org, Retrieved 3 July, 2019
- Fish-shrimps, fishery: tnau.ac.in, Retrieved 23 May, 2019

Chapter 4

Harvesting Algae and Seaweed

Algaculture is a type of aquaculture that deals with the farming of species of algae. Based on their size, algae are classified as microalgae and macroalgae. Algae cultivation is used to produce various products such as algae biodiesel, bioplastics, omega-3 fatty acids etc. This chapter has been carefully written to provide an easy understanding of these varied facets of harvesting algae and seaweed.

Algaculture

Algaculture is the commercial cultivation of algae. Algae are simple green plants that grow in water. Their green color means they produce their own food using photosynthesis, just like grass, trees and corn. Algae come in two main forms. Macroalgae are seaweeds. Kelp grows to more than 180 feet (55 meters) long in the ocean. Nori is the variety you'll find wrapped around your sushi. Microalgae are tiny, single-celled plants that float in the water, each one visible only through a microscope.

Algaculture is nothing new. Seaweed was first cultivated in Japan at least 1,500 years ago and algae production is still a big business there. Dulse has long been eaten in the British Isles and the microalgae spirulina were harvested by the Aztecs of 16th-century Mexico. In addition to providing human food, seaweeds have been used for fertilizers. They provide the food thickener carrageen and other gelling agents and stabilizers that show up in everything from soup to toothpaste. Worldwide, algae production is a $6 billion business.

Today, algae are attracting new interest and resarch investment because of their potential to provide energy and combat environmental threats. Part of the organic mass of algae takes the form of oil, which can be squeezed out and converted to biodiesel fuel. Algae beat land plants hands down in the efficiency with which they produce oil. Some varieties of algae yield an oil that can be refined into gasoline and even jet fuel. The carbohydrate portion of the plants can be fermented for ethanol production.

Algae can convert waste carbon dioxide, a greenhouse gas that pours from smokestacks, to usable products. They can help clean dirty water, converting pollutants to biomass. They have additional uses in pharmaceuticals and cosmetics.

The Promise of Algae

Why have algae generated excitement and attracted research investment in recent years? Like all green plants, algae contain chloroplasts in their cells. These tiny structures are charged with chlorophyll, a molecule that uses light energy to combine carbon and water into a simple sugar. The cells further transform some of these sugars into proteins and lipids or oil.

But if algae are doing the same thing as corn, wheat and apple trees, why bother raising them? After all, corn on the cob, sweet rolls and apple pie taste better than seaweed to most of us. Here are some of the things algae have going for them:

- Productivity: Algae are super fast-growing. Land plants take months or years to reach maturity. Algae can complete their entire life cycle in a single day. Some algae can double their biomass in just an hour.
- Efficiency: When it comes to converting solar energy to biomass, algae are all business. Because they're supported by and take their nutrients directly from water, they need no roots, stems or flowers. Land plants use as much as 95 percent of their energy building the structures they need for support, feeding and breeding.
- Concentration: Because of their efficiency, algae can be grown in a very concentrated space. They produce up to 100 times more oil per acre than land plants.
- Versatility: It's estimated that there are more than 70,000 species of algae, many of them not yet classified. Growers can pick ones suited to conditions and goals, selecting varieties for a specific temperature range or water salinity.
- Non-competition: Algae don't compete with current crops for land or fresh water. They can be grown in ponds in locations, like deserts, that don't sustain land plants. Some varieties prefer saline or polluted water.

Attracted by all these advantages, algae cultivators have been working diligently to come up with efficient and economical ways to grow and harvest the plants. The cost factor is currently the great challenge that must be overcome to make algae commercially viable.

Commercial Cultivation of Algae

All algaculture requires three basics water, light and nutrients.

Water's the easiest. It doesn't need to be potable; different types of algae grow nicely in fresh water, salt water and dirty water. Sunlight, because it's free, is the preferred light. But sunlight reaches only 3 or 4 inches (7 to 10 centimeters) into a mass of algae, so growers must agitate the algae to expose all of it to the light. The main nutrient is carbon dioxide, which can come from the air or other source. Agitation or bubbling dissolves it into the water. The grower must supply other nutrients, like nitrogen and trace elements, if they aren't already in the water.

The growth of algae generally takes place in one of three different forms:

- Open Pond/Raceway.
- Tubular Bioreactor.
- Vertical Bag Bioreactor.

Open Pond/Raceway

A raceway pond grows algae as it moves along a water channel circuit that is usually about 30 cm deep. The base of the winding channel is usually constructed out of compacted dirt and/or cement and lined with a white reflective material. Flow through the channel is provided by a paddle-wheel that is located between the extractor of mature algae and the feeder of new algae, while the flow is directed around the bends by baffles, as shown in the Raceway ponds are desirable for their large scalability and low set-up costs but experience significant water loss through evaporation. The raceway pond also experiences temperature fluctuations and carbon dioxide losses to the surrounding environment. Productivity is affected by contamination with unwanted algae types and microorganisms that feed on algae. The concentration of algae can remain low due to raceways being poorly mixed.

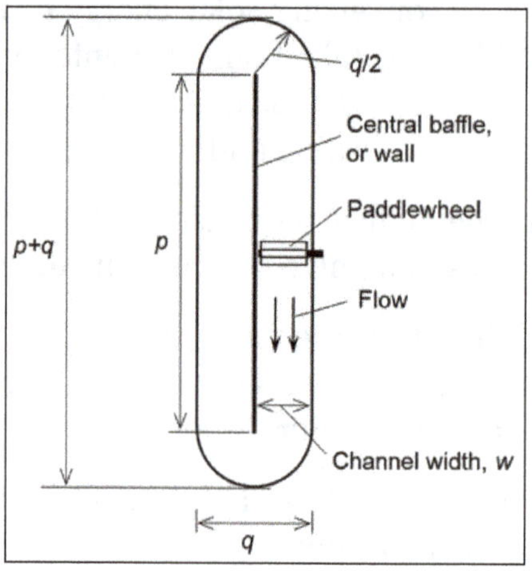

Open Pond/Raceway

Tubular Bioreactor

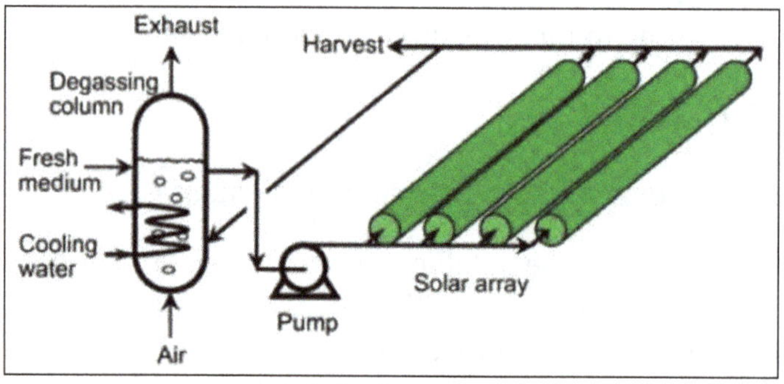

Tubular bioreactor

A tubular bioreactor grows algae in an array of clear plastic or glass tubes that have CO_2 constantly fed through the closed system along with the necessary nutrients, such as nitrogen, phosphorus, and iron. The tubes are generally less than 10cm in diameter to allow for light to penetrate to all of the algae. The array of tubes is arranged horizontally, vertically, or on a slant but always oriented running north to south. This arrangement allows for the semi-continuous growth of algae, as the fresh algae culture is being fed into the bioreactor at the same rate at which the mature algae is being drawn out. Algae production is stopped periodically in order to clean the tubes. The recycled growth medium and fresh young culture in a new medium are degassed to remove the oxygen and bubbles from the new culture before being fed into the tubular bioreactor. Excess oxygen needs to be removed from the system because oxygen concentrations above 4 times that of air will prevent algae growth. The feeding of CO_2 and other inorganic supplements mentioned above happens at the entrance to the tubes and at locations along the length of the tube as required. Testing sites are necessary along the length of the tubular bioreactor to control the pH and inorganic salt concentrations at different stages of the algae growth. The best photosynthetic conversion rates are achieved at tubular diameters of 2-5 cm.

Vertical Bag Bioreactor

A vertical bag system grows algae in an aqueous solution within a closed bag usually 2-4 cm thick for optimum photosynthetic efficiency. The CO_2 and other inorganic nutrients are fed through the system from the top, bottom, or sides of the bag and bubbled through the algae. Exhaust gases rich in oxygen are siphoned off of the top of the vertical bags to reduce the oxygen in the growth medium, as too much oxygen inhibits algae growth. The vertical bag system was derived from the tubular bioreactor system in order to reduce material start-up costs and allow for easier modification of the system (adding and removing testing and feeding points is much easier with a vertical bag).

The vertical bag system is very good for a laboratory setting because algal batches of any size can be made with it. Lab trials have found that the bags needed to be replaced every 60-90 days as contaminants build up in the system and prevent effective algae growth.

Vertical bag bioreactor

All of these systems are designed for growing microalgae, the one-celled varieties that float in water. Growers usually cultivate macroalgae in the open sea. The water already contains the nutrients the algae need and provides good growing conditions. The traditional method was simply to harvest wild seaweed, and this is still done in coastal areas around the world.

With increased demand, growers began to cultivate seaweed. For some varieties, such as kelp, spores are attached to ropes that are then anchored in the ocean and the seaweed is allowed to grow. Other types grow from pieces of seaweed that are fixed to nets or deposited in pools.

Agriculture has been around for 10,000 years. Algaculture is relatively new. Scientists and engineers are actively studying the best ways to raise algae efficiently. The harvesting of plants is another subject of intense research.

Harvesting and Processing Algae

Harvesting microalgae means removing the microscopic plants from the water in which they grow and concentrating them into a paste. The grower then needs to remove the moisture, leaving a dense biomass. The minute size of microalgae presents a problem when it comes to harvesting.

1. One method is filtration: The grower can run the water containing the algae through a cellulose membrane whose pores are smaller than the algae cells. This can be difficult because filters quickly fill up with algae and become clogged. Researchers are looking for better ways to efficiently filter algae.

A worker stacks baskets of harvested algae at the waterfront.

2. Flocculation, another method of harvest, means getting the algae to clump together. Adding chemicals or types of algae that naturally clump can cause microalgae to form clumps that become easier to gather.

3. Another way to harvest algae is by flotation. Here, the grower uses compressed air to create a froth of bubbles and algae that brings the tiny plants to the surface where they can be skimmed off.

4. A centrifuge is yet another harvest method. Spinning a container filled with water and algae causes the algae to collect in one end.

5. In order to harvest their crops most effectively, algaculture growers sometimes combine these methods. They might use flocculation to form algae clumps, then separate them with flotation or a centrifuge. Coming up with a truly efficient way to harvest microalgae is a key to bringing down the cost of cultivation.

6. Harvesting macroalgae involves different problems. Gathering wild seaweed is a labor-intensive process. Some types of seaweed grown in controlled conditions can be gathered in nets. Kelp raised on ropes can be hauled out and hung up to dry. Kelp forests in shallow seas can be mowed by machines, taking off the tops of undersea kelp beds.

7. Once harvested, algae must be drained of its water and dried. A centrifuge can spin water out, but is relatively expensive. Some systems combine harvest and processing, spreading the algae on belt filters that let the water drain through, then removing more water using a capillary medium that draws water out of the biomass of algae.

8. The next step is to break down the cell walls of the algae in order to extract the oil inside. The algae are put through a screw or piston press. Chemicals, electromagnetic pulses or ultrasound may also be used to break down the cells. When the oil has been drained off, the remaining biomass is compressed into a cake to be used as to supplement animal feed or as a fertilizer. Algae have found a wide range of uses, the most exciting ones in the energy field.

The Many uses for Algae

The buzz about algae is that it's an ideal source of renewable energy and could be the ultimate green fuel. Research by the U.S. government and companies like Boeing, Chevron and Honeywell are developing ways to make algaculture an economically viable foundation for a new generation of energy. Part of the attraction is the range of fuels into which algae can be converted:

- Biodiesel is the simplest way to tap algae's energy potential. Like any vegetable oil, oil from algae can be chemically transformed into biodiesel fuel. Compared to land plants like soybeans or corn, algae use less land and fresh water, grow faster and have higher concentrations of oil.
- Refined transportation fuels are another area of promise for algae. Some algae produce oil that can be refined into gasoline or even jet fuel, and without the sulfur and nitrogen compounds in petroleum. Manufacturers can process it in the same refineries as petroleum-based stock. In 2011, the first commercial jet flight powered by algal oil flew from Houston to Chicago.
- Ethanol, which is commonly added to gasoline, can be produced from algae as well as land plants. Besides oil, algae are made up from carbohydrates and cellulose walls. These materials can be fermented by yeast into ethanol or grain alcohol.
- Methane, the main ingredient in natural gas, is produced when bacteria digest algae. A clean and versatile fuel, methane can be used to produce electricity or power vehicles. It represents another biofuel option for algae.

Algae actually thrive on polluted water, which means they can be used for waste water treatment. Algae turn pollutants from municipal, industrial or agricultural waste water into usable byproducts like animal feed or biomass for conversion to energy. Algae naturally accumulate heavy metals for removal or recycling.

Because carbon dioxide, the greenhouse gas that contributes to climate change, is algae's favorite food, the plants can be used for carbon capture. They convert the gas to organic carbon compounds at a far faster clip than land plants. One pound (453.6 grams) of algae consumes 2 pounds (907.2 grams) of carbon dioxide. Feed the waste gas of a coal-burning power plant into a mass of algae, and they literally eat it up. Waste gas can be stored for permanent elimination from the atmosphere, or converted to fuel to cut the use of fossil fuels.

Algae continue play a role as human food and supplements. People eat seaweed in salads and sushi and take supplements made from the microalgae spirulina. Algae provide complete protein, omega-3 fatty acids and vitamins. Carageen is extracted from red seaweed known as Irish moss and used as a thickener.

Algae are also being used as feed for cattle and for marine animals like shrimp and shellfish. The biomass left after algae have been processed can sometimes be applied as organic fertilizer to farm fields. Algae find minor uses in cosmetics and pharmaceuticals as well.

Research into growing, harvesting and processing algae is advancing on many fronts. Given its immense value, there's no doubt that this simple "weed" will play a growing role in the future of our society and economy.

Algae Biodiesel

The Potential of Algae Oil as a Fuel Source

Over 80% of the energy we use comes from three fossil fuels: petroleum, coal, and natural gas. About 98% of carbon emissions result from fossil fuel combustion. About 98% of carbon emissions result from fossil fuel combustion. Reducing the use of fossil fuels would significantly reduce the amount of carbon dioxide and other pollutants produced. This can be achieved by either using less energy altogether or by replacing fossil fuel by renewable fuels. Renewable energy is a promising alternative solution because it fixes CO_2 in the atmosphere through photosynthesis. They also produce lower or negligible levels of greenhouse gases and other pollutants when compared with the fossil energy sources they replace.

Table: Chemical composition of algae on a dry matter basis (%).

Species of sample	Proteins	Carbohydrates	Lipids	Nucleic acid
Scenedesmus obliquus	50-56	10-17	12-14	3-6
Scenedesmus quadricauda	47	-	1.9	-
Scenedesmus dimorphus	8-18	21-52	16-40	-
Chlamydomonas rheinhardii	48	17	21	-
Chlorella vulgaris	51-58	12-17	14-22	1-5
Chlorella pyrenoidosa	57	26	2	-
Spirogyra sp.	6-20	33-64	11-21	-
Dunaliella bioculata	49	4	8	-
Dunaliella salina	57	32	6	-
Euglena gracilis	39-61	14-18	14-20	-
Prymnesium parvum	28-45	25-33	22-38	1-2
Tetraselmis maculate	52	15	3	-
Porphyridium	28-39	40-57	9-14	-
Spirulina platensis	46-63	8-14	4-9	2-5
Spirulina maxima	60-71	13-16	6-7	3-4.5
Synechoccus sp.	63	15	11	5
Anabaena cylindrica	43-56	25-30	4-7	-

Algae, like corn, soybeans, sugar cane, wood, and other plants, use photosynthesis to convert solar energy into chemical energy. They store this energy in the form of oils, carbohydrates, and proteins. The plant oil can be converted to biodiesel, which is why biodiesel is a form of solar energy. The more efficient a particular plant is at converting that solar energy into chemical energy, the better it is from a biodiesel perspective, and algae are among the most photosynthetically efficient plants on earth.

The annual productivity and oil content of algae is far greater than seed crops. Soybean can only produce about 450 l of oil per hectare. Canola can produce 1200 l per hectare, and palm can produce

6000 l. Algae, on the other hand, can yield 90,000 l per hectare. Microalgae contain lipids and fatty acids as membrane components, storage products, metabolites and sources of energy. Algae contain anywhere between 2% and 40% of lipids/oils by weight. Algae can grow anywhere there is enough sunshine and some can grow in saline water. All algae contain proteins, carbohydrates, lipids and nucleic acids in varying proportions. Microalgae can complete an entire growth cycle every few days. Although the percentages may vary, there are types of algae that are comprised up to 40% of their overall fatty acids. The culture of algae can yield 30-50% oil. Oil supply is based on claims that 47,000-308,000 l/hectare/year of oil could be produced using algae.

Highest-yielding Algae

The algae used in biodiesel production are usually aquatic unicellular green algae. This type of algae is a photosynthetic eukaryote characterized by high growth rates and high population densities. Under good conditions, green algae can double its biomass in less than 24 hours. Green algae can also have high lipid contents, usually over 50%. This high yield is ideal for intensive agriculture and can be an excellent source for biodiesel production.

Table: Oil contents of some microalgae strains

Microalgae	Oil content (wt% of dry basis)
Botryococcus braunii	25-75
Chlorella sp.	28-32
Cryphecodinium cohnii	20
Cylindrotheca sp.	16-37
Dunaliella primolecta	23
Isochrysis sp.	25-33
Monallanthus salina	>20
Nannochloris sp.	20-35
Nannochloropsis sp.	31-68
Neochloris oleoabundans	35-54
Natzschia sp.	45-47
Phaeodactylum tricornutum	20-30
Schizochytrium sp.	50-77
Tetraselmis sueica	15-23

Chlorella

Chlorella is a single-celled green algae belonging to the class of Chlorophyceae. It is a primary algae because it grows autotrophically. It is spherical in shape, about 2 to 10 μm in diameter, and does not have a flagella. Chlorella has green photosynthetic pigments, chlorophyll-a and chlorophyll-b, in its chloroplast. Using photosynthesis, it multiplies rapidly requiring only carbon dioxide, water, sunlight, and a small amount of minerals to reproduce. Chlorella is believed to be capable in serving as a potential food and energy source because of its photosynthetic efficiency to reach 8% comparable to other highly efficient crops such as sugar cane.

Manipulation of metabolic pathways can redirect cellular function toward the synthesis of preferred products and even increase the processing capabilities of microalgae. For heterotrophic microalgae, outside carbon sources offer a large amount chemical energy, which the cells often store as lipid droplets. Heterotrophically cultivated Chlorella protothecoides has been shown to accumulate as much as 55% of its dry weight as oil, compared to only 14% in cells grown photoautotrohpically.

Table: Yield of various plant oils.

Crop	Oil in liters per hectare
Algae	1,00,000
Castor	1413
Coconut	2683
Palm	5950
Safflower	779
Soy	446
Sunflower	952

Dunaliella

Dunaliella is a unicellular green algae also belonging to the class of Chlorophyceae. It too is a primary algae. It is rod to oval shaped and about 9 to 11 μm in diameter. The organisms are simple to cultivate and do not clump or form chains.

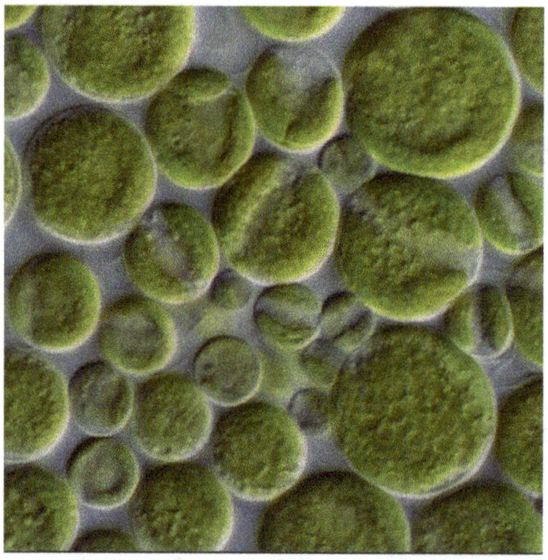

Light microscopy of Chlorella protothecoides.

The properties of various fatty esters determine the overall fuel properties of the biodiesel fuel. There is no one strain or species of algae that can be said to be the best in terms of oil yield for biodiesel. But, diatoms along with green algae are the most promising. Scenedesmus dimorphus is a unicellular algae in the class Chlorophyceae (green algae). While this is one preferred species for oil yield for biodiesel, one of the problems with Scenedesmus is that it is heavy, and forms thick sediments if not kept in constant agitation. Dunaliella tertiolecta is a marine green flagellate with a size of 10 to 12 μm in diameter. This strain is said to have an oil yield of about 37%. D. tertiolecta is a fast growing strain, therefore allowing it to have a high carbon dioxide rate.

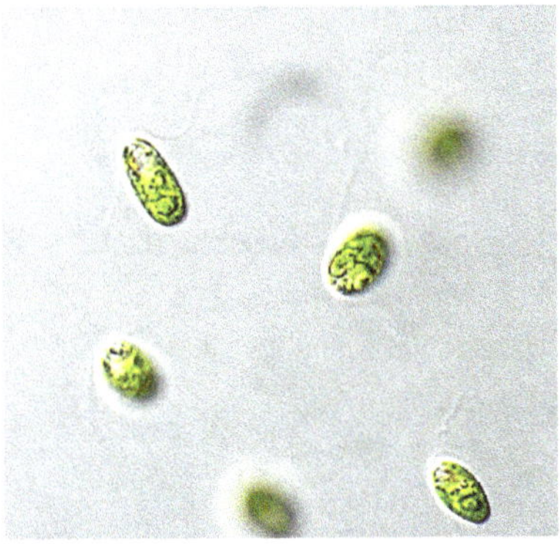

Light microscopy of Dunaliella tertiolecta.

Processes Converting Algae Oil to Biodiesel

Photosynthesis is the first step in the conversion of light to chemical energy and ultimately responsible for supporting all biofuel synthetic processes, converting solar energy into biomass, carbon storage products (carbohydrates and lipids) and hydrogen.

In green algae, light is captured by light harvesting complex proteins, referred to as LHCI and LHCII. Their expression is dependent on environment conditions (light intensity). These proteins bind a large amount of the chlorophyll and carotenoids in the plant and play a role in both light capture and in the dissipation of excess energy, which would otherwise inhibit the photosynthetic reaction centers (photosystem II). Excitation energy used to drive the photosynthetic reactions is funneled to the photosynthetic reaction centers of photosystem I (PSI) and photosystem II (PSII) via the network of pigments bound by the LHC, PSII, and PSI subunits. In the beginning, PSII uses this energy to drive the photosynthetic water splitting reaction, which converts water into protons, electrons and oxygen. The electrons are passed along the photosynthetic electron transport chain via plastoquinone (PQ), cytochrome b_6f ($Cyt\, b_6f$), PSI and ferredoxin (Fd) and on to NADPH. At the same time, protons are released into the thylakoid lumen by PSII and the PQ/PQH_2 cycle. This generates a proton gradient, which drives ATP production via ATP synthase. The protons and electrons are recombined by ferredoxin-$NADP^+$ oxidoreductase (FNR) to produce NADPH. NADPH and ATP are used in the Calvin cycle to produce the sugars, starch, oils that are required to produce bioethanol, biodiesel, and biomethane.

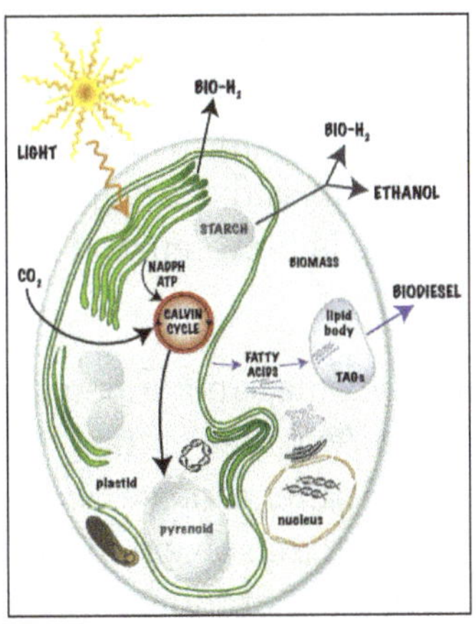

Metabolic pathways in microalgae related to biofuel production. The integration of metabolic pathways is coordinated through complex mechanisms that regulate photosynthetic output for the synthesis of proteins, nucleic acids, carbohydrates, lipids, and H_2.

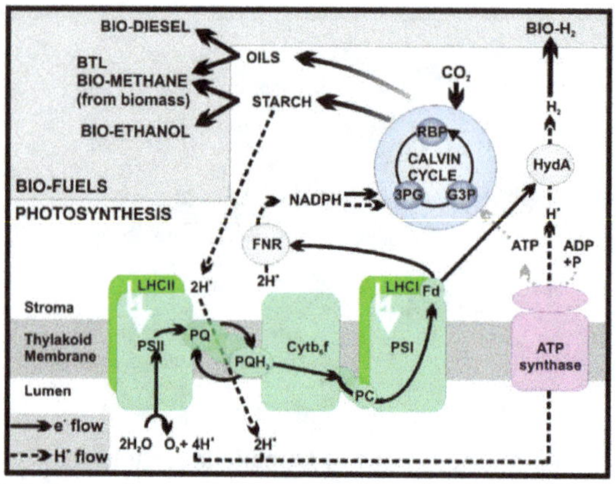

The process of photosynthesis converts solar energy into chemical energy and is key to all biofuel production systems in microalgae.

The Calvin cycle is an integral part of the photosynthetic process and responsible for fixing CO_2 in a diverse range of organisms including primitive algae through to higher plants. The process uses ATP and NADPH generated by the light reactions.

Algal oil is converted into biodiesel through a transesterification process. Oil extracted from the algae is mixed with alcohol and an acid or a base to produce the fatty acid methylesters that makes up the biodiesel. An excess of methanol is used to force the reaction to favor the right side of the equation. The excess methanol is later recovered and reused. If biomass is grown in a sustained way, its combustion has no impact on the CO_2 balance in the atmosphere, because the CO_2 emitted by the burning of biomass is offset by the CO_2 fixed by photosynthesis.

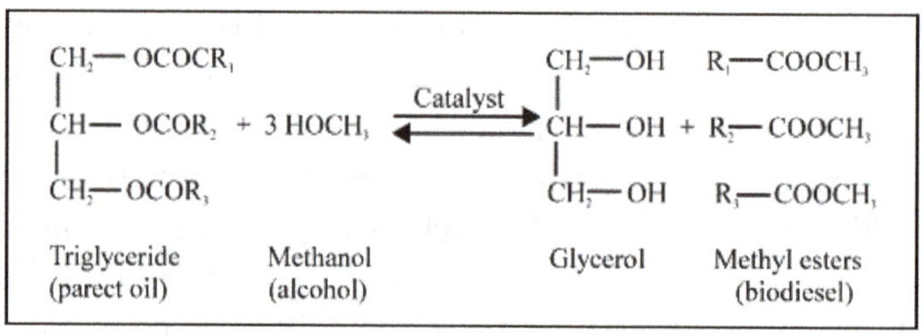

Transesterification chemical equation.

The process of producing microalgal oil consists of producing microalgal biomass that requires light, carbon dioxide, water and inorganic nutrients (nitrates, phosphates, and iron). About half of the dry weight of microalgal biomass is carbon, which is usually derived from carbon dioxide. Therefore, producing 100 tons of algal biomass fixes roughly 183 tons of carbon dioxide. Optimal temperature for growing many microalgae is between 293 and 303 K. A temperature outside this range could kill or damage the cells.

Expeller/Press, solvent extraction with hexane, and supercritical fluid extraction are well-known methods to extract oil from algae. A press/expeller extracts 70-75% of the oils out of algae. Using chemicals like Hexane (which are relatively inexpensive) can also be used to extract algal oils. Supercritical fluid extraction is more efficient than solvent separation methods. Because supercritical fluids are selective, they provide high purity and product concentrations and can extract nearly 100% of the oils. In the supercritical fluid (CO_2) extraction, CO_2 is liquefied under pressure and heated to the point that it has the properties of both a liquid and gas. This liquefied fluid then acts as the solvent in extracting the oil. After oil extraction from algae, the remaining biomass fraction can be used as a high protein feed for livestock. This gives further value to the process and reduces waste.

Biology and Adaptation

Microalgae grow quickly and contain high oil content compared with terrestrial crops, which take a season to grow and only contain a maximum of about 5 percent dry weight of oil. They commonly double in size every 24 hours. During the peak growth phase, some microalgae can double every three and one-half hours. Oil content of microalgae is usually between 20 percent and 50 percent, while some strains can reach as high as 80 percent. This is why microalgae are the focus in the algae-to-biofuel arena.

Table: Oil content of microalgae.

Microalga	Oil content (% dry weight)
Botryococcus braunii	25-75
Chlorella sp.	28-32
Crypthecodinium cohnii	20
Cylindrotheca sp.	16-37
Nitzschia sp.	45-47
Phaeodactylum tricornutum	20-30
Schizochytrium sp.	50-77
Tetraselmis suecia	15-23

Table: Oil yields based on crop type.

Crop	Oil yield (gallons/acre)
Corn	18
Soybeans	48
Canola	127
Jatropha	202
Coconut	287
Oil Palm	636
Microalgae	6283-14641

Production and Agronomic Information

Most microalgae are strictly photosynthetic — that is, they need a light and carbon dioxide as energy and carbon sources. This culture mode is usually called photoautotrophic. Some algae species, however, are capable of growing in darkness and using organic carbons such as glucose or acetate as energy and carbon sources. This culture mode is termed heterotrophic. Due to high capital and operational costs, heterotrophic algal culture is hard to justify for biodiesel production. In order to minimize costs, algal biofuel production usually relies on photoautotrophic culture that uses sunlight as a free source of light.

Phototrophic microalgae require light, carbon dioxide, water, and inorganic salts to grow. The culture temperature should be between 15 and 30 °C (~60-80 °F) for optimal growth. The growth medium must contribute the inorganic elements that help make up the algal cell, such as nitrogen, phosphorus, iron, and sometimes silicon. For large-scale production of microalgae, algal cells are continuously mixed to prevent the algal biomass from settling and nutrients are provided during daylight hours when the algae are reproducing. However, up to one-quarter of algal biomass produced during the day can be lost through respiration during the night.

A variety of photoautotrophic-based microalgal culture systems are available. For example, the algae can be grown in suspension or attached on solid surface. Each system has its own advantages and disadvantages. Currently, the suspension-based open ponds and enclosed photobioreactors are commonly used for algal biofuel production. In general, an open pond is simply a series of raceways outside, while a photobioreactor is a sophisticated reactor design which can be placed indoors in a greenhouse, or outdoors.

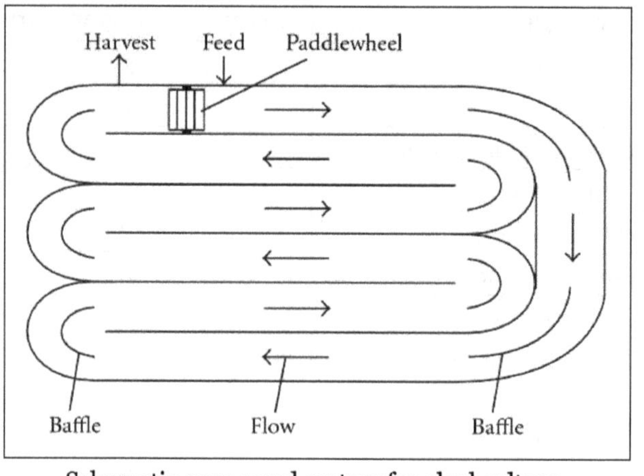

Schematic open pond system for algal culture.

The details of the two systems are described below:

1. Open ponds: Open ponds are the oldest and simplest systems for mass cultivation of microalgae. In this system, the shallow pond is usually about 1 foot deep; algae are cultured under conditions identical to their natural environment. The pond is designed in a raceway configuration, in which a paddlewheel provides circulation and mixing of the algal cells and nutrients. The raceways are typically made from poured concrete, or they are simply dug into the earth and lined with plastic to prevent the ground from soaking up the liquid. Baffles in the channel guide the flow around bends in order to minimize space. The system is often operated in a continuous mode – that is, the fresh feed containing nutrients including nitrogen phosphorus and inorganic salts is added in front of the paddle wheel. Algal broth is harvested behind the paddle wheel after it has circulated through the loop. Depending on the nutrients required by algal species, a variety of wastewater sources can be used for the algal culture, such as dairy/swine lagoon effluent and municipal wastewater. For some marine types of microalgae, seawater or water with high salinity can be used.

Although open ponds cost less to build and operate than enclosed photobioreactors, this culture system has its intrinsic disadvantages. Since these are open-air systems, they often experience a lot of water loss due to evaporation. Thus, microalgae growing in an open pond do not uptake carbon dioxide efficiently, and algal biomass production is limited. Biomass productivity is also limited by contamination with unwanted algal species as well as other organisms from feed. In addition, optimal culture conditions are difficult to maintain in open ponds, and recovering the biomass from such a dilute culture is expensive.

2. Enclosed photobioreactors: Enclosed photobioreactors have been employed to overcome the contamination and evaporation problems encountered in open ponds. These systems are made of transparent materials and generally placed outdoors for illumination by natural light. The cultivation vessels have a large surface area-to-volume ratio.

The most widely used photobioreactor is a tubular design, which has a number of clear transparent tubes, usually aligned with the sun rays. The tubes are generally less than 10 centimeters in diameter to maximize sunlight penetration. The medium broth is circulated through a pump to the tubes, where it is exposed to light for photosynthesis, and then back to a reservoir. The algal biomass is prevented from settling by maintaining a highly turbulent flow within the reactor, using either a mechanical pump or an airlift pump. A portion of the algae is usually harvested after the solar collection tubes. In this way, continuous algal culture is possible. In some photobioreactors, the tubes are coiled spirals to form what is known as a helical tubular photobioreactor, but these sometimes require artificial illumination, which adds to the production cost. Therefore, this technology is only used for high-value products, not biodiesel feedstock.

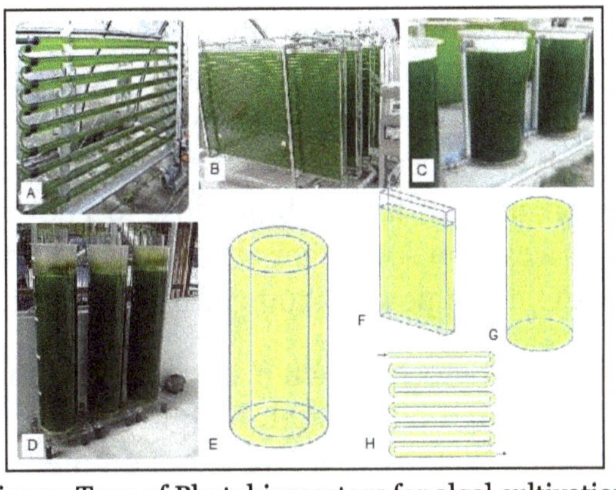

Figure: Type of Photobioreactors for algal cultivation

The photosynthesis process generates oxygen. In an open-raceway system, this is not a problem as the oxygen is simply returned to the atmosphere. However, in the closed photobioreactor, the oxygen levels will build up until they inhibit and poison the algae. The culture must periodically be returned to a degassing zone, an area where the algal broth is bubbled with air to remove the excess oxygen. Also, the algae use carbon dioxide, which can cause carbon starvation and an increase in pH. Therefore, carbon dioxide must be fed into the system in order to successfully cultivate the microalgae on a large scale. Photobioreactors may require cooling during daylight hours, and the temperature must be regulated at night hours as well. This may be done through heat exchangers, located either in the tubes themselves or in the degassing column.

The advantages of the enclosed photobioreactors are obvious. They can overcome the problems of contamination and evaporation encountered in open ponds. The biomass productivity of photobioreactors can be 13 times greater than that of a traditional raceway pond, on average. Harvesting of biomass from photobioreactors is less expensive than that from a raceway pond, since the typical algal biomass is about 30 times as concentrated as the biomass found in raceways. However, enclosed photobioreactors also have some disadvantages. For example, the reactors are more expensive and difficult to scale up. Moreover, light limitation cannot be entirely overcome since light penetration is inversely proportional to the cell concentration. Attachment of cells to the tube walls may also prevent light penetration. Although enclosed systems can enhance the biomass concentration, the growth of microalgae is still suboptimal due to variations in temperature and light intensity.

Harvesting: After growing in open ponds or photobioreactors, the microalgae biomass needs to be harvested for further processing. The commonly used harvest method is through gravity settlement, or centrifuge. The oil from the biomass will be removed through solvent extraction and further processed into biodiesel.

Environmental and Sustainability Issues

In addition to producing biofuel, algae can also be explored for a variety of other uses, such as fertilizer and pollution control. Certain species of algae can be land-applied for use as an organic fertilizer, either in its raw or semi-decomposed form. Algae can be grown in ponds to collect fertilizer runoff from farms; the nutrient-rich algae can then be collected and reapplied as fertilizer, potentially reducing crop-production costs. In wastewater-treatment facilities, microalgae can be used to reduce the amount of chemicals needed to clean and purify water.

In addition, algae can also be used for reducing the emissions of CO_2 from power plants. Coal is, by far, the largest fossil energy resource available in the world. About one-fourth of the world's coal reserves reside in the United States. Consumption of coal will continue to grow over the coming decades, both in the United States and the world. Through photosynthetic metabolism, microalgae absorb CO_2 and release oxygen. If an algae farm is built close to a power plant, CO_2 produced by the power plant could be utilized as a carbon source for algal growth, and the carbon emissions would be reduced by recycling waste CO_2 from power plants into clean-burning biodiesel.

Microalgae are an ideal biodiesel feedstock, which eventually could replace petroleum-based fuel due to several advantages, such as high oil content, high rates of production, less land, etc. Currently, algal biodiesel production is still too expensive to be commercialized. Due to the static costs associated with oil extraction and biodiesel processing and the variability of algal biomass production, cost-saving efforts for algal oil production should focus on the production method of the oil-rich algae itself. This needs to be approached through enhancing both algal biology (in terms of biomass yield and oil content) and culture-system engineering. In addition, using all aspects of the microalgae for producing various value-added products besides the algal fuel, via an integrated biorefinery, is an appealing way to lower the cost of algal biofuel production. Indeed, microalgae contain a large percentage of oil, with the remaining parts consisting of large quantities of proteins, carbohydrates, and other nutrients. This makes the post-oil extraction residue attractive for use as animal feed or in other value-added products.

Seaweed

Seaweed isn't used to describe a certain species - it's a common name for a variety of types of plants and plant-like creatures, from tiny phytoplankton to enormous giant kelp. Some seaweeds are true, flowering plants (an example of these are seagrasses). Some aren't plants at all but are algae, which are simple, chloroplast-containing organisms that don't have roots or leaves. Like plants, algae do photosynthesis, which produces oxygen.

The algae shown here have pneumatocysts, which are gas-filled floats that allow the blades of the seaweed to float towards the surface. Why is this important? This way the algae can reach the sunlight, which is crucial for photosynthesis.

Algae are classified into three groups: red, brown, and green algae. While some algae have root-like structures called holdfasts, algae do not have true roots or leaves. Like plants, they do photosynthesis, but unlike plants, they are single-celled. These single cells may exist individually or in colonies. Initially, algae were classified in the plant kingdom. Classification of algae is still under debate. Algae are often classified as protists, eukaryotic organisms that have cells with a nucleus, but other algae are classified in different kingdoms. An example is blue-green algae, which are classified as bacteria in the Kingdom Monera.

Phytoplankton are tiny algae that float in the water column. These organisms lie at the foundation of the ocean food web. Not only do they produce oxygen through photosynthesis, but they provide food for countless species of other marine life. Diatoms, which are yellow-green algae, are an example of phytoplankton. These provide a food source for zooplankton, bivalves (e.g., clams) and other species.

Plants are multi-cellular organisms in the kingdom Plantae. Plants have cells that are differentiated into roots, trunks/stems and leaves. They are vascular organisms that are capable of moving fluids throughout the plant. Examples of marine plants include seagrasses (sometimes referred to as seaweeds) and mangroves.

Seagrasses

Seagrasses like those shown here are flowering plants, called angiosperms. They live in marine or brackish environments worldwide. Seagrasses are also commonly called seaweeds. The word seagrass is a general term for about 50 species of true seagrass plants.

Seagrasses need lots of light, so they are found at relatively shallow depths. Here they provide food for animals such as the dugong, along with shelter for animals such as fish and invertebrates.

Uses of Seaweeds

The use of seaweed as food has been traced back to the fourth century in Japan and the sixth century in China. Today those two countries and the Republic of Korea are the largest consumers of seaweed as food. However, as nationals from these countries have migrated to other parts of the world, the demand for seaweed for food has followed them, as, for example, in some parts of the United States of America and South America. Increasing demand over the last fifty years outstripped the ability to supply requirements from natural (wild) stocks. Research into the life cycles of these seaweeds has led to the development of cultivation industries that now produce more than 90 percent of the market's demand. In Ireland, Iceland and Nova Scotia (Canada), a different type of seaweed has traditionally been eaten, and this market is being developed. Some government and commercial organizations in France have been promoting seaweeds for restaurant and domestic use, with some success. An informal market exists among coastal dwellers in some developing countries where there has been a tradition of using fresh seaweeds as vegetables and in salads.

China is the largest producer of edible seaweeds, harvesting about 5 million wet tonnes. The greater part of this is for kombu, produced from hundreds of hectares of the brown seaweed, Laminaria japonica, that is grown on suspended ropes in the ocean. The Republic of Korea grows about 800 000 wet tonnes of three different species, and about 50 percent of this is for wakame, produced from a different brown seaweed, Undaria pinnatifida, grown in a similar fashion to Laminaria in

China. Japanese production is around 600,000 wet tonnes and 75 percent of this is for nori, the thin dark seaweed wrapped around a rice ball in sushi. Nori is produced from a red seaweed - a species of Porphyra. It is a high value product, about US$ 16,000/dry tonne, compared to kombu at US$ 2,800/dry tonne and wakame at US$ 6,900/dry tonne.

Various red and brown seaweeds are used to produce three hydrocolloids: agar, alginate and carrageenan. A hydrocolloid is a non-crystalline substance with very large molecules and which dissolves in water to give a thickened (viscous) solution. Alginate, agar and carrageenan are water-soluble carbohydrates that are used to thicken (increase the viscosity of) aqueous solutions, to form gels (jellies) of varying degrees of firmness, to form water-soluble films, and to stabilize some products, such as ice cream (they inhibit the formation of large ice crystals so that the ice cream can retain a smooth texture).

Seaweeds as a source of these hydrocolloids dates back to 1658, when the gelling properties of agar, extracted with hot water from a red seaweed, were first discovered in Japan. Extracts of Irish Moss, another red seaweed, contain carrageenan and were popular as thickening agents in the nineteenth century. It was not until the 1930s that extracts of brown seaweeds, containing alginate, were produced commercially and sold as thickening and gelling agents. Industrial uses of seaweed extracts expanded rapidly after the Second World War, but were sometimes limited by the availability of raw materials. Once again, research into life cycles has led to the development of cultivation industries that now supply a high proportion of the raw material for some hydrocolloids.

Today, approximately 1 million tonnes of wet seaweed are harvested and extracted to produce the above three hydrocolloids. Total hydrocolloid production is about 55 000 tonnes, with a value of US$ 585 million.

Alginate production (US$ 213 million) is by extraction from brown seaweeds, all of which are harvested from the wild; cultivation of brown seaweeds is too expensive to provide raw material for industrial uses.

Agar production (US$ 132 million) is principally from two types of red seaweed, one of which has been cultivated since the 1960-70s, but on a much larger scale since 1990, and this has allowed the expansion of the agar industry.

Carrageenan production (US$ 240 million) was originally dependent on wild seaweeds, especially Irish Moss, a small seaweed growing in cold waters, with a limited resource base. However, since the early 1970s the industry has expanded rapidly because of the availability of other carrageenan-containing seaweeds that have been successfully cultivated in warm-water countries with low labour costs. Today, most of the seaweed used for carrageenan production comes from cultivation, although there is still some demand for Irish Moss and some other wild species from South America.

Seaweed meal, used an additive to animal feed, has been produced in Norway, where its production was pioneered in the 1960s. It is made from brown seaweeds that are collected, dried and milled. Drying is usually by oil-fired furnaces, so costs are affected by crude oil prices. Approximately 50 000 tonnes of wet seaweed are harvested annually to yield 10,000 tonnes of seaweed meal, which is sold for US$ 5 million.

Fertilizer uses of seaweed date back at least to the nineteenth century. Early usage was by coastal dwellers, who collected storm-cast seaweed, usually large brown seaweeds, and dug it into local soils. The high fibre content of the seaweed acts as a soil conditioner and assists moisture retention, while the mineral content is a useful fertilizer and source of trace elements. In the early twentieth century, a small industry developed based on the drying and milling of mainly storm-cast material, but it dwindled with the advent of synthetic chemical fertilizers. Today, with the rising popularity of organic farming, there has been some revival of the industry, but not yet on a large scale; the combined costs of drying and transportation have confined usage to sunnier climates where the buyers are not too distant from the coast.

The growth area in seaweed fertilizers is in the production of liquid seaweed extracts. These can be produced in concentrated form for dilution by the user. Several can be applied directly onto plants or they can watered in, around the root areas. There have been several scientific studies that prove these products can be effective. In 1991, it was estimated that about 10,000 tonnes of wet seaweed were used to make 1,000 tonnes of seaweed extracts with a value of US$ 5 million. However, the market has probably doubled in the last decade because of the wider recognition of the usefulness of the products and the increasing popularity of organic farming, where they are especially effective in the growing of vegetables and some fruits.

Cosmetic products, such as creams and lotions, sometimes show on their labels that the contents include "marine extract", "extract of alga", "seaweed extract" or similar. Usually this means that one of the hydrocolloids extracted from seaweed has been added. Alginate or carrageenan could improve the skin moisture retention properties of the product. Pastes of seaweed, made by cold grinding or freeze crushing, are used in thalassotherapy, where they are applied to the person's body and then warmed under infrared radiation. This treatment, in conjunction with seawater hydrotherapy, is said to provide relief for rheumatism and osteoporosis.

Over the last twenty years there have been some large projects that investigated the possible use of seaweeds as an indirect source of fuel. The idea was to grow large quantities of seaweed in the ocean and then ferment this biomass to generate methane gas for use as a fuel. The results showed the need for more research and development, that it is a long-term project and is not economic at present.

There are potential uses for seaweed in wastewater treatment. Some seaweeds are able to absorb heavy metal ions such as zinc and cadmium from polluted water. The effluent water from fish farms usually contains high levels of waste that can cause problems to other aquatic life in adjacent waters. Seaweeds can often use much of this waste material as nutrient, so trials have been undertaken to farm seaweed in areas adjacent to fish farms.

Sources of Seaweed

Brown Seaweeds

The main uses of brown seaweeds are as foods and as the raw material for the extraction of the hydrocolloid, alginate. The more useful brown seaweeds grow in cold waters in both the Northern and Southern Hemispheres. They thrive best in waters up to about 20 °C. Brown seaweeds are found in warmer waters, but these are less suitable for alginate production and rarely used as food.

Brown Seaweeds as Food

Food from brown seaweeds comes mostly from the genera Laminaria, Undaria and Hizikia. Originally, harvests of wild seaweeds were the only source, but since the mid-twentieth century demand has gradually outstripped the supply from natural resources and methods for cultivation have been developed. Today, seaweed for food comes mainly from farming rather than natural sources.

Species of the genus Laminaria are eaten in Japan and China, and to a lesser extent in the Republic of Korea. Laminaria was native to Japan and the Republic of Korea, and was introduced accidentally to China, in 1927 at the northern city of Dalian, probably by shipping. Prior to that, China had imported its needs from the naturally growing resources in Japan and the Republic of Korea. In the 1950s, China developed a way of cultivating Laminaria on long ropes suspended in the ocean, and this became a widespread source of income for large numbers of coastal families. By 1981, they were producing 1,200,000 wet tonnes annually of seaweed. In the late 1980s, production fell as some farmers switched to the more lucrative but risky farming of shrimp. By the mid-1990s, production had started to rise and the reported harvest in 1999 was 4,500,000 wet tonnes. China is now self-sufficient in Laminaria and has a good export market.

Laminaria was in plentiful supply in Japan, mainly from the northern island of Hokkaido, where several naturally growing species were available. However, as Japan increasingly prospered after the Second World War, demand grew, and by the 1970s cultivation became necessary. They now draw their supply from a mixture of natural and cultivated harvests. In the Republic of Korea, the demand for Laminaria is much lower and most is now provided from cultivation.

Undaria has been harvested from natural resources for many years in the Republic of Korea, China and Japan. The Republic of Korea has the highest consumption of the three countries. Cultivation commenced in the Republic of Korea and Japan in the 1960s, but not until the mid-1980s in China. By 1999, the Republic of Korea was producing about 5,000 wet tonnes of wild seaweed and about 250,000 wet tonnes by cultivation. Some of this was exported to Japan, where production was only 3,000 wet tonnes of wild harvest and 77,000 wet tonnes by cultivation. Laminaria is more popular than Undaria in China, and by the mid-1990s China was harvesting about 100,000 wet tonnes of Undaria from cultivation, small by comparison with the 3 million wet tonnes of Laminaria at that time.

Hizikia is popular in Japan and the Republic of Korea. It has been harvested from natural beds, up to 20,000 wet tonnes in the Republic of Korea in 1984, when cultivation was commenced. Since then cultivation, on the southwest coast of the Republic of Korea, has steadily increased so that by 1994 about 32,000 wet tonnes were farmed and only 6,000 wet tonnes harvested from the wild. A large proportion of the Republic of Korea production is exported to Japan, where there is little activity in cultivation of this species.

Alginate-containing Seaweeds

These are called alginophytes - needing only one word instead of three to describe the seaweed.

These are nearly all harvested from natural resources. A wide variety of species are used, harvested in both the Northern and Southern Hemispheres. Countries include Argentina, Australia, Canada, Chile, Ireland, Norway, Mexico, South Africa, United Kingdom (Scotland and Northern

Ireland) and United States of America. Cultivation of brown seaweeds like Laminaria and Undaria go through the sexual reproduction cycle, a time consuming and labour intensive process that is expensive, even in low-labour-cost countries. Cultivated raw material is normally too expensive for alginate production. While much of the Laminaria that is cultivated in China is used for food, when there is surplus production this can be used in the alginate industry, probably provided at a lower price that is subsidized by the high price obtained on the food market.

Red Seaweeds

The main uses of red seaweeds are as food and as sources of two hydrocolloids: agar and carrageenan. Useful red seaweeds are found in cold waters such as Nova Scotia (Canada) and southern Chile; in more temperate waters, such as the coasts of Morocco and Portugal; and in tropical waters, such as Indonesia and the Philippines.

Red Seaweeds as Food

Porphyra species are the largest source of food from red seaweeds. Porphyra, known by the more common names of nori and laver, is dried and processed into thin purplish-black sheets. One of its common uses is in Japanese sushi, where it is wrapped on the outside of a small handful of soured, boiled rice topped with a piece of raw fish. Porphyra has been cultivated in Japan and the Republic of Korea since the seventeenth century; there are natural stocks, but even at that time they were insufficient to meet demand. Cultivation was developed intuitively, by observing the seasonal appearance of spores, but Porphyra has a complex life cycle that was not understood until the 1950s. Since then, cultivation has flourished, and today the supply is virtually all from cultivation, which is conducted on a large scale in Japan, China and the Republic of Korea. In 1999, the combined production from these three countries was just over 1,000,000 wet tonnes. It has the highest value of any cultivated seaweed, about US$ 1,200 per wet tonne. For comparison, the brown seaweeds used as food are valued at US$ 610/wet tonne for Laminaria and US$ 530/ wet tonne for Undaria.

Dulse (Palmaria palmata, formerly Rhodymenia palmata) is another red seaweed used as food, but on a very small scale, mostly collected by coastal people from natural resources and consumed locally. It is dried and sold in whole pieces, usually eaten without cooking, or as a powder that is used as a condiment. It grows in cold waters and is collected in Ireland, Iceland and the east coast of Canada. There is a small industry in the Bay of Fundy (Canada), and a company based in Belfast, Northern Ireland (United Kingdom), is working to expand the market for dulse.

Agar-containing Seaweeds (Agarophytes)

Two genera, Gelidium and Gracilaria, account for most of the raw material used for the extraction of agar. Extraction of Gelidium species gives the higher quality agar (as measured by the gel strength: the strength of a jelly formed by a 1.5 percent solution).

All Gelidium used for commercial agar extraction comes from natural resources, principally from France, Indonesia, the Republic of Korea, Mexico, Morocco, Portugal and Spain. Gelidium is a small, slow growing plant and while efforts to cultivate it in tanks/ponds have been biologically successful, it has generally proved to be uneconomic. However, one company, Marine BioProducts

International, has launched high-grade agar and agarose products that they claim are derived from their own cultivated Gelidium. Presumably the profit from these products at the high end of the market is sufficient to offset the costs of cultivation. Small quantities of Gracilariopsis are harvested in Chile and species of Gelidiella provide the raw material for a small agar industry in India.

Gracilaria species were once considered unsuitable for agar production because the quality of the agar was poor (gel strength too low). In the 1950s, it was found that pre-treatment of the seaweed with alkali before extraction lowered the yield but gave a good quality agar. This allowed expansion of the agar industry, previously limited by the supply of Gelidium available, and led to the harvesting of a variety of wild species of Gracilaria in countries such as Argentina, Chile, Indonesia and Namibia. Chilean Gracilaria was especially useful, but soon there was evidence of overharvesting of the wild crop. Cultivation methods were then developed, both in ponds and in the open waters of protected bays. These methods have spread beyond Chile to other countries, such as China, the Republic of Korea, Indonesia, Namibia, the Philippines and Viet Nam, usually using species of Gracilaria native to each particular country. Obviously, Gracilaria species can be grown in both cold and warm waters. Today the supply of Gracilaria still comes mainly from the wild, with the degree of cultivation depending on price fluctuations.

Carrageenan-containing Seaweeds (Carrageenophytes)

The original source of carrageenan was the red seaweed Chondrus crispus (common name: Irish Moss), collected from natural resources in France, Ireland, Portugal, Spain and the east coast provinces of Canada. As the carrageenan industry expanded, the demand for raw material began to strain the supply from natural resources, although by this time Chondrus was being supplemented by species of Iridaea from Chile and Gigartina from Spain.

The introduction of cultivation of species of Eucheuma in the Philippines during the 1970s provided the carrageenan industry with a much enhanced supply of raw material. A further advantage of this cultivated material was that one species contained almost exclusively a particular type of carrageenan (kappa-carrageenan) while a second species contained predominantly a second type (iota-carrageenan), each type having its own particular applications. Chondrus contains a mixture of two types (kappa and lambda) that could not be separated during commercial extraction. Today most of the raw material comes from the two species originally cultivated in the Philippines, but their cultivation has now spread to some other warm-water countries, such as Indonesia and Tanzania. Limited quantities of wild Chondrus are still used; attempts to cultivate Chondrus in tanks have been successful biologically, but uneconomic as a raw material for carrageenan. Wild species of Gigartina and Iridaea from Chile are also being harvested and efforts are being made to find cultivation methods for these.

The two species originally cultivated in the Philippines were named Eucheuma cottonii and Eucheuma spinosum, and the industry shortened these so they are often referred to as "cottonii" and "spinosum". However, botanists have since renamed both species, so that Eucheuma cottonii is now Kappaphycus alvarezii, while Eucheuma spinosum is now Eucheuma denticulatum. Unfortunately all these names are still in use, so an awareness of them is necessary when reading about carrageenophytes.

Seaweed Farming

The culture of the tropical seaweed species Kappaphycus alvarezii and Eucheuma denticulatum began in the 1960s in the Philippines, with commercial-scale production reached in 1971. Cottonii and Spinosum, respectively, are the commercial names of these two species. Since those early days, global seaweed aquaculture production was reached 180,000mt dry-weight of Cottoni, and 28,000mt dry-weight of Spinosum. presents summary of world production.

Table: World production figure (mt dry-weight).

Country of Production	2001 (mt/yr)	2002 (mt/yr)	2003 (mt/yr)	2009 (mt/yr) (est'd.)	2010 (mt/yr (est'd.)
Philippines a) Cottonii b) Spinosum	118,400 4,100	125,200 5,500	127,700 3,000		80,000 20,000
Indonesia a) Cottonii b) Spinosum	28,800 3,460	35,600 3800 mt	36,800 3,100		90,000 5,000
Tanzania (Zanzibar) a) Cottonii b) Spinosum	742 2,480	921 2,561	1,378 4,600		500 12,000
Malaysia a) Cottonii	3,200	3,600	4,900		3,000
Vietnam a) Cottonii					4,000
Madagascar a) Cottonii b) Spinosum				2009 Estimated 1,500 500	1,000 1,000
Total a) Cottonii b) Spinosum	151,142 10,040	165,321 11,861	170,778 10,700		178,500 38,000

In the WIO-region, seaweed culture was initiated in the 1980s in Tanzania, particularly on the islands of Zanzibar and Pemba, and it was subsequently taken-up in Madagascar and Mozambique in the mid-1990s. Kenya first introduced seaweed culture, albeit in a commercially unsuccessful trial, in 2004 and then again in 2009 with the support of an EU-COI project, with the first container-loadshipped in 2011. However, these are not in fact the only countries where an introduction has been attempted, but only a few countries have been able to develop an industrial-production of at least 1,000mt dry-weight, and then maintain it over more than a few years.

In general, Spinosum is mainly grown in Tanzania and Kenya, with Cottoni the dominant species in Madagascar.

Uses of Carrageen as an Extract

The principle interest in seaweed is for the extraction of carrageenans, which are processed into gels that are subsequently used in the food industry, comestics, and pharamaceuticals as thickeners, stabilising agents and emlusifiers. One of the important characteristics of carrageenan-based gels is that they remain fluid under pressure and then recover their original viscosity.

The following are some example of specific uses of carrageenan gels:

- Food Industry: deserts, ice-creams, concentrated milk, pasta, processed meats, sauces and chinese soups, in beermaking processes, soy milk, animal feed, dietetic drinks, jams etc.
- Comestics: toothpaste, shampoo, skin-care creams etc.
- Pharamceuticals: in pills, gels etc.
- Others: for example, in fire-extinguishers and polish.

Introduction to Seaweed Farming

Seaweeds aregenerally cultivated in lagoons or sheltered bays.Seaweeds obtain nutrients directly from the seawater and it is crucial to have currents that flush the site in which the seaweeds are placed. Small cuttings are attached to a line with short lengths of string (known as 'Tie Tie') at a density of five cuttings per metre of line. In nature these seaweeds live attached to a solid substrate.

Spinosum

Spinosum is a robust seaweed and is relatively easy to cultivate because it the best suited to withstanding temperature variations and infestation by algal parasites.

Spinosum on the coast of southern Kenya.

Cottonii

Cottonii is more difficult to cultivate and requires more care and more skill. It is a seaweed that is very sensitive to infestation by algal parasites and to water temperatures above 31-32 °C. And a farm or village, even one that is technically well supported, may see much of its Cottonii seaweed disappear once or twice per year. This annual loss requires a production cycle to be established specifically to produce new cuttings.

A harvest of Cottonii being transported by outrigger canoe in SW Madagascar.

Daily Growth Rate

The growth rate of seaweed located in a suitable habitat, even taking into account seasonal variations, is impressive. A single cutting of 100g can reach a weight of 1kg in 20-40 days, which could yield at least 8 seperate harvests per year.

The increase in weight of a 100g cutting according to a range of daily growth rates.

% croissancepar jour// growth rate per day	3% 100 gr	4% 100 gr	5% 100 gr	6% 100 gr	7% 100 gr	8% 100 gr
Nombre de jours// Number of days						
10 j	134 gr	148 gr	163 gr	179 gr	197 gr	216 gr
20 j	181	219	265	321	387	466
30 j	243	324	432	574	761	1006
45 j	378	584	899	1376	2100	3192

In summary, one of the keys to the culture of seaweed is to have a minimum daily growth rate of 5-8% over the majority of the year. It is advised not to start a seaweed farming venture if the average growth rate through the 12-months is less than 4%.

The growth rate of 4% takes account of the unavoidable negative growth rate in one or two months each year due to infestation by parasitic algae, grazing fish, the disease known as Ice-ice, high sea temperature, storms and periodic inflows of fresh water. The impact of these negative events may be further exacerbated by a farmer's lack of management of the seaweed plots.

Techniques of Seaweed Culture

Off-bottom (Post and Line)

The general approach is to suspend a series of lines of 10m in length between two posts, which are usually made of wood. This technique is best suited for lagoons, where there is relatively shallow water at low tide and for small-scale initiatives. The lines are regularly checked and the seaweed harvested during the two spring tide periods each month, at which time the farmers can work on foot, and is therefore a technique particularly suited to women farmers. Following each harvest, the seaweed is dried on land for a few days before sale.

Introduction of Spinsoum farming using the off-bottom method.

Longlines

This technique involves the use of line of up to 50m in length, anchored at each end and with floats attached every 10m or so to support the line This technique is usually employed in water of between 4 and 10m and farmers therefore require access to some sort of boat to access the plots. However, as a result of the necessity of a boat the farmers can access the plots at all times, except for during bad weather.

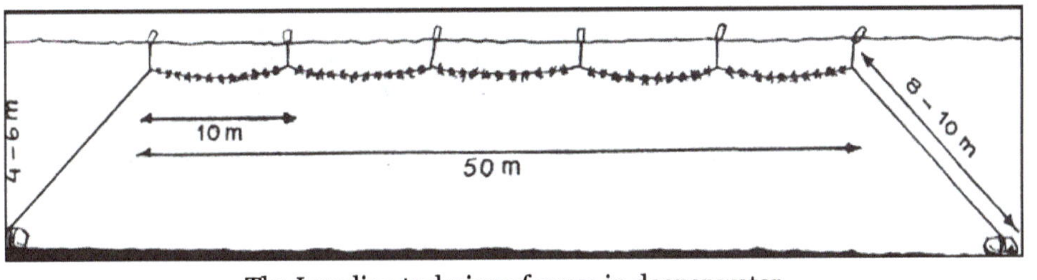

The Longline technique for use in deeper water.

Rock-based Farming for Spinosum

Rock-based farming originated in Asia and was first introduced into the WIO region, in Zanzibar, under a EU-COI project. As the name suggests, cuttings of Spinosum are initially attached to a rock (roughly the size of a large fist) with an elastic band, although after a few weeks the seaweed esablishes its own fixation points. The ideal density of rocks is 25 per m^2.

With this technique the harvesting of seaweed is done,at low tide and on foot, by simply cutting away the new growth each cycle while leaving enough of the 'root stock' for the cycle of growth to start over again. This technique avoids the need to afix new cuttings after each harvest and represents a significant labour-saving; for example, a farmer using the line-based techniques would require up to 120,000 individual harvest manipulations per year to manage lines of just 3kms (= 15,000 seperate plants). This technique has obvious attractions and an image given by farmers using this technique in Zanzibar was that is was like having an orchard of mangos, that provide their fruit all year-round. However, there is one major constraint to the application of this technique and that is the requirement for a site wellprotected from potentially rough sea conditions. In sites that are too exposed it is a certitude that the rocks will end up on the beach or at the bottom of the lagoon.

Rocks that are naturally seeded with Spinosum.

A Spinosum farm in Uzi Village, Zanzibar, using rocks with seaweed cuttings attached. At a density of 25 rocks per square-metre and 1.5 million rocks laid by 100 families, total production should reach 1,000mt per year.

Floating Rafts

This technique is employed principally in protected bays in Asia.

The Village/Family Approach

It is very much at the family and village-level where the environmental and socio-economic charateristicsfavour seaweed farming. After all, the work unit is usually the family and their reward is the sale of dried seaweed to the buyers. Operating in a mutually respectful but nevertheless under a contractual arrangement, the basic nature of the relationship between these two partners can be characterised as follows:

- Buyers Farmers: The buyers provide the farmers with the raw materials for production, including the lines, sometimes the posts, the boats, and the drying tables. They also link farmers with technical support and provide supervision. All this support is provided under contract with clear rules that help to weed-out the less-motivated farmers and to reallocate support to others. Payments are made in cash at least twice monthly.

- Farmers Buyers: It is essential that farmers respect the exclusive nature of theircontracts,which requires the sale of their seaweed to the contract partner only. It is also essential that the necessary quality of dry seaweed is produced.

The Industrial Approach

In the industrial approach, an investor would look for a favourable site and engage workers who would be expected to work for 22-days per month. The investor would provide close and continuous supervision and the level of wages paid are directly linked to the level of production of seaweed. This approach is predicated on the availability of individuals who are motivated by the need for a cash income; and they are often temporary economic migrants seeking to earn just enough money for their own 'projects'. It is important that any plan to use immigrant labour is carefully introduced accepted by the local resident population. The presence of an industrial set-up does not imply that villagelevel seaweed farms cannot co-exist in the same area.

For private-sector investors, the advantages of an industrial/commercial operation compared to family and village based farming are:

- A higher quality of labour, resulting in a better management of disease and fewer unforeseen problems.
- A better management of the quality of the final product with a concomitant expectation of a higher final price.
- Better protection against 'pirate' buyers. For example, Tanzania has witnessed an invasion of Chinese buyers who pay a higher beach price than farmers could expect under their contracts. But the Chinese buyers have usually invested nothing in developing production capacity. And they can further improve their profit margins by saving money on transport by shipping seaweed back to China in empty containers that had previously brought merchandise from China to Tanzania. This is an unsustainable approach because it ultimately discourages the very private investment that created the seaweed farms in the first place.

The major constraint of the medium-scale commercial approach is that one replaces a relatively simple system of farming and harvesting based on family units (even if there are, for example, 300 families operating across five villages) with a SME of 120 employees. Furthermore, it is often necessary therefore to provide accomodation for the workers and ensure adequate provision of general supplies and foodstuffs for the site in general.

Technical, socio-economic and financial challenges and the means to address them:

- Aide Memoire: Key Facts about Seaweed Farming.
    - Cottonii is much more vulnerable to environmental factors than is the case with Spinosum, and therefore it is harder to farm it successfully;
    - From a technical perspective, seaweed that is under environmental stress may at best stop growing, and at worst, display what is in effect a negative growth rate as plant biomass is lost;

- Cottonii sells at a price at least double that of Spinosum;
- The sector of course depends on both producers and buyers, but for the latter a financial break-even production is reached at 300mt/yr for Cottonii and 500mt/year for Spinosum;
- To reach financial break-even requires three years at least;
- International buyers are not interested in exporters that cannot provide a least 1,000mt dry-weight per year;
- Evidence suggests that a private-sector investor has only about a 25% of succeeding in the seaweed farming business.

The Disease Ice-Ice

Ice-Ice develops as a result of stresses on the seaweed. The tell-tale signs of the disease is that the thalli (or frond) degenerate, turn a pale white colour similar to ice (hence the name of the disease) and eventually rot. It is a clear sign that seaweed is not healthy and it eventually literally falls apart with the consequential loss of a production cycle in a matter of a few days. Perhaps more serious is the loss of the supply of cuttings for the following production cycle.

A seaweed plant with Ice-Ice.

Water Temperature

Sea-water temperature is also a limiting factor on growth, and should ideally be between 20° et 32 °C, while a sudden change in temperature (even it f it remains within the ideal range) can also have a negative effect on seaweed growth.

Prolonged Exposure of Seaweed to the Air (Especially for Cottonii)

Exposure of seaweed to the air can have a chronic effect on growth rates, even if it doesn't necessarily result in the full-blown development of Ice Ice. This problem is associated with farmers who systematically take new cuttings only when the harvest has been brought ashore for cleaning and drying, rather than taking cuttings in situ. A farmer can lose the equivalent weight of one or two individual harvests per year, perhaps without even realising it, because the harvest cycle can be1-12 days longer than what it should ideally be.

The taking of cuttings in situ is important to avoid stressing the seaweed and reducing its growth rate

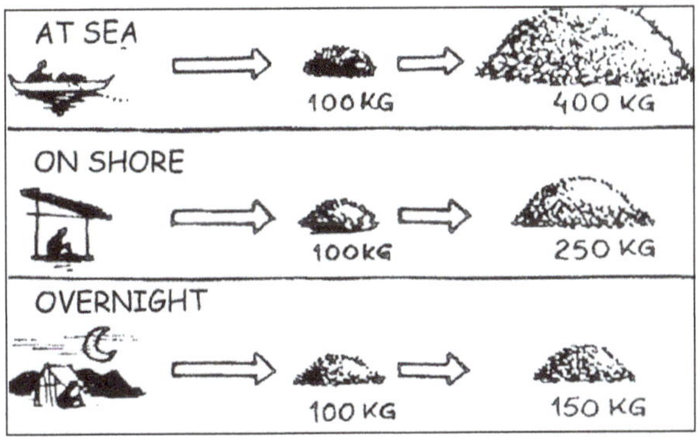

On the basis of a growth-rate of 4.5% per day for 30-days, this figure indicates the general difference in production as a function of exposure to the air. In fact, seaweed that under environmental stress can stop growing for between 1 and 12 days. The farmer therefore loses, without even knowing it, one or two harvests per year.

Salinity

Salinity is also a limiting factor on seaweed growth, particularly if it drops rapidly. Such a rapid drop in salinity would usually occur as a result of heavy rains in adjacent river catchment areas. Overall it is necessary to avoid siteswhere salinity is outside of the range of 23-38 grammes/litre (ppm); the average salinity in the western Indian Ocean is 35- 36ppm. Freshwater is more buoyant than seawater and one solution to the risk of freshwater influx is to set the lines at least 1-metre below the sea surface.

Algal Parasites (Cottoni is more Vulnerable than Spinosum)

To best understand the nature of the problem of algal parasites of seaweed the following is an example of what happened to a well-managed and productive Cottoni farm in NE Madagascar.

The company had attained a production of Cottoni of close to 2,000 mt dry-weight per year. Then a massive infestation of Epiphytic Filamentous Algae (EFA) took place that caused a 50% decline in production in 2010. This, exacerbated by the presence of other parastic algal, discouraged the farmers to the extent that production had almost ceased in 2011.

The infestation of seaweed plots by EFA (or 'pest' seaweed) is recurrent but often seasonal. It can bring on amassive Ice Ice event, where some or even allof the production can be lost. And farmers may then be left without stock suitable for cuttings and are therefore obliged to spend one harvest cycle for no other reason than to replace this stock. It is likely, due to pest seaweed, that two of the normal eight harvests per year will be lost.

How to Limit the Damage

- Build on past experience;
- Ensure good monitoring of the farm itself, particularly during the seasons where algal parasites are more common. In fact, the infestation by EFA develops progressively, permitting time for a response;
- Mobilise the technicians to support the best farmers;
- Increase the diversity of the locations of the seaweed plots;
- Diversify the modes of production. In fact, longlines set deeper in the water column will be less frequently and less rapidly infested than seaweeds gowing off-bottom in the lagoon.

Kenyan technician checking the lines of a non-motivated farmer (no farmer's shadow on the seaweed). In such situations, if no improvements are seen, it is advisable to reduce the size of the individual's farm and ultimately to end the contract.

- Ensure a rapid mobilisation of the farmers to harvest the seaweed because seaweed harvested and dried is equivalent to money in the bank and prevents farmers becoming discouraged;
- Attach only small cuttings (e.g. cuttings of 25-50gr as opposed to the normal 100gr) with the remaining healthy seaweed;
- Establish a back-up nursery in a site that is not contaminated by the infestion;
- Mobilise the farmers to clean the seaweed still in situ on the re-stocked lines.

The six most damaging parasitic algal are the following:

- Macro-algae: Enteromorpha, Ulva, Chaetomorpha, Hypnea and Hydroclathus.

Spinosumparasitised by Enteromorpha.

- Epiphytic Filamentous Algae (EFA).

Cottoniparasitised by EFA in a commercial farm in the north-east of Madagascar. This farm saw its Cottonii production decline by 50% in 2010, while in 2011 it was almost zero.

Turbidity

Seaweeds can endure a certain level of turbidity. But they are vulnerable to high turbidity caused by terrigenous material, that can stick to the thalli and induce Ice Ice. Farmers need to regularly shake their lines to remove the silty deposits from the seaweed.

Storms and Cyclones

It is obvious that seaweed farms are vulnerable to bad weather. To minimise its impact on the farm, it is advised to chose sites that are sheltered whereever possible. Other than that, it is important to monitor weather forecasts, and if the forecast is threatening then famers should mobilise and harvest the available seaweed to avoid any loss, and replace the stock with smaller cuttings (of 25g).

Grazing of Seaweed

By far the biggest losses from grazing of the seaweed crop is to herbivorous fish (often Rabbitfish (Siganidae)) although normally this is only a seasonal problem. These fish usually shelter around rocks and coral bommies and therefore the only mitigation available is to avoid locating seaweed plots close to such features. The longline technique, which sets lines a few metres below the surface, is less vulnerable. Grazing represents a real but relatively minor loss of production to plots of mature seaweed, but nurseries can be more vulnerable and suffer greater losses.

The Best Strains of Seaweed

In the WIO, all the seaweed farming is of varieties (or strains) of just two species, Kappaphycus alvarezii (Cottonii) and Eucheuma denticulatum (Spinosum). Different strains of seaweed do not share the same biological charateristics. Some strains are better adapted to, for example, higher water temperaturs but at the same time may have lower growth rates. All the strains currently cultivated in the WIO originate from The Philipines and Indonesia because their growth rates are better than the native strains.

If investors or farmers wish to introduce new strains and/or species then an Environmental Impact Assessment (EIA) will be neccesary. In a recent example from Kenya, the EIA process took 18-months to complete. In general, diversifying the species/strains used for seaweed production is not straightforward and requires full participation of buyers and the relevent Government ministries, as well as a significant level of competence.

Socio-Economic Issues

In making the choice of sites, as well as needing suitable environmental conditions for the growth of seaweed, the socio-economic characteristics of the local population and their ability and willingness to work on the farms are essential to consider. What are their needs? What other marketable resources are available? Are individuals and families willing to work hard? And what are the main characteristics (e.g. relating to power relations, wealth distribution, ethnic mix etc.) of the village? These are some of the many questions that are better dealt with before investment in farming begins, rather than after.

Post-harvest Treatment of Seaweed

The Quality of the Dry Seaweed

Drying is managed at the level of the primary producers, the farmers, and it is essential that the techniques employed by them result in the required quality standard (30% moisture content, no damage from freshwater and the absence of sand and other foreign bodies) and the required sanitary standard (e.g. no fecal material from domestic animals). These strict requirements demand active quality control by the technicians in charge of buying, including a mechanism for tracing seaweed back to the specific farmer, as well as a series of final checks before the purchased seaweed is pressed into bales.

Storage

The seaweed bales should be stored in a some sort of 100% wateright shed or warehouse. It is not recommended to store bales of seaweed for more than 6-months and they must be kept dry to avoid the onset of fermentation that could degrade the quality of the carrageenans.

The Seaweed Press

The purchase of a press (~ US$5-7,000) should be seriously considered; they can sometimes be obtained second-hand from sisal or cotton mills. But a suitable press can also be built locally. In fact, it is best to export the seaweed as quickly as possible to maintain a cash-flow for the enterprise. And a twenty-foot (6m) container is able to hold 20-22mt dry-weight of seaweed.

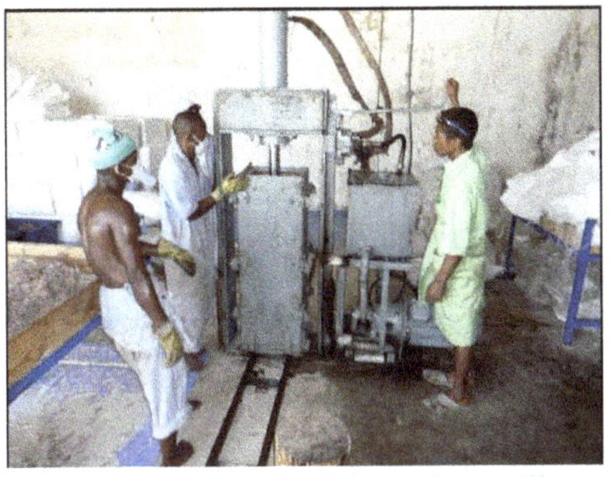

A seaweed press built in Madagascar.

Planning for Export

One would expect to be able to fill the first container after 12-18 months and the second container three to six months later. The objective is to reach a production level of one container per month within 21/2 years.

References

- Algaculture, biofuels, fuel-efficiency: howstuffworks.com, Retrieved 10 February, 2019
- Growing-algae: energyeducation.ca, Retrieved 21 August, 2019
- Biodiesel-from-Algae-Oil: kenyon.edu, Retrieved 17 April, 2019
- Algae-for-biofuel-production: extension.org, Retrieved 8 January, 2019
- What-is-seaweed: thoughtco.com, Retrieved 29 March, 2019

Chapter 5

Integrated Multi-Trophic Aquaculture

Integrated multi-trophic aquaculture is a practice in which the inputs for one species are derived from the recycled by-products from another species. It promotes economic and environmental sustainability. The diverse applications of integrated multi-trophic aquaculture in the current scenario have been thoroughly discussed in this chapter.

Aquaculture sometimes called mariculture when in the sea – is the production of fish, invertebrates (e.g. bivalves) and plants (seaweeds) in aquatic systems and by a variety of production methods. Typically these different types of organisms are grown separately. Integrated Multi Trophic Aquaculture (IMTA) is a concept where different species are grown together in such a way that the invertebrates and/or plants can recycle the nutrients that are lost from the culture of the other species.

An example of this might be a fish farmer who adds fish feed to his cages of fish in the sea knowing that a proportion of the feed, all of the faeces and most of the nutrients in that feed will not end up in his fish but will be lost to the environment. However, if the farmer develops a bivalve culture operation near his fish farm, these filter-feeders might benefit by consuming some of the particles of feed and faeces and so grow faster or bigger than they might otherwise have done. Additionally the farmer might elect to grow some seaweed near his fish farm. These plants can utilise the nitrogen excreted by the fish and the bivalves to enhance their growth. The result is that there is a net reduction of losses to the environment plus new crops to harvest and sell.

To continue to grow, while developing better management practices, the aquaculture sector needs to develop more innovative, responsible, sustainable, and profitable technologies and practices, hich should be ecologically efficient, environmentally benign, product-diversified, and societally eneficial. Maintaining sustainability, not only from an environmental, but also from economic, social, and technical perspectives, has become a key issue, increased by the enhanced awareness of more and more demanding consumers regarding quality, traceability, and production conditions.Integrated multi-trophic aquaculture (IMTA) has the potential to play a role in reaching these objectives by cultivating fed species (e.g., finfish or shrimps fed sustainable commercial diets) with extractive species, which utilize the inorganic (e.g.,seaweeds or other aquatic vegetation) and rganic (e.g., suspension- and deposit-feeders) excess nutrients fromaquaculture for their growth.

Conceptual diagram of an Integrated Multi-Trophic Aquaculture (IMTA) operation including the combination of fed aquaculture (e.g., finfish) with suspension organic extractive aquaculture (e.g., shellfish), taking advantage of the enrichment in small particulate organic matter (POM); inorganic extractive aquaculture (e.g., seaweeds), taking advantage of the enrichment in dissolved inorganic nutrients (DIN); and deposit organic extractive aquaculture (e.g., echinoids, holothuroids, and polychaetes), taking advantage of the enrichment in large particulate organic matter (POM) and feces and pseudo-feces (F&PF) from suspension-feeding organisms. The bioturbation on the bottom also regenerates some DIN, which becomes available to the seaweeds.

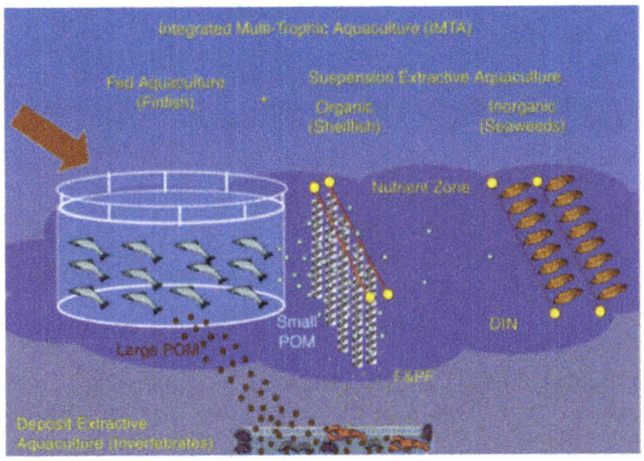

IMTA: A Flexible and Functional Concept

The IMTA concept is extremely flexible. To use a musicology analogy, IMTA is the central/overarching theme on whichmany variations can be developed according to the environmental, biological, physical, chemical, societal, and economic conditions prevailing in parts of the world where the IMTA systems are operating. It can be applied to open-water or land-based systems, marine or freshwater systems (sometimes called "aquaponics"), and temperate or tropical systems. What is important is that the appropriate organisms are chosen at multiple trophic levels based on the complementary functions they have in the ecosystem, as well as for their economic value or potential. In fact, IMTA is doing nothing other than recreating a simplified, cultivated ecosystem in balance with its surrounding instead of introducing a biomass of a single type one thinks can be cultivated in isolation from everything else.

Integration should be understood as cultivationin proximity, not considering absolute distances but connectivity in terms of ecosystemic function a lities. It should be made clear that in the minds of those who created the acronym "IMTA," it was never conceived to be viewedwith the minimalist perspective of only the cultivation of salmon (Salmo salar), kelps (Saccharina latissima and Alaria esculenta), and blue mussel s (My tilusedulis) within a few hundred meters: this is only one of the variations and the IMTA concept can be extended to very large ecosystems like the Yellow Sea. This also means that IMTA variations include integrated agriculture aquaculture systems (IAAS), integrated sylviculture (mangrove) aquaculture systems (ISiAS), integrated green water aquaculture systems (IGWAS), integrated peri-urban aquaculture systems (IPUAS), integrated fisheries aquaculture systems (IFAS), sustainable ecological aquaculture systems (SEAS), integrated temporal aquaculture systems (ITAS), and integrated sequential aquaculture systems (ISAS, also called partitioned aquaculture systems, PAS, or fractionated aquaculture systems, FAS). There is no ultimate IMTA system to "feed the world." There is not one world but climatic, environmental, biological, physical, chemical, economic, societal, and political conditions, each of which can lead to different choices of systems for feeding these microworlds.

The paradox is that IMTA is not a new concept. Asian countries, which provide more than two thirds of the world's aquaculture production, have been practicing IMTA (often described as a type of "polyculture") for centuries, through trial and error and experimentation. Why, then, is this common sense solution not more widely implemented? The reasons for this generally center around social customs and practices, and market-driven economic models not considering externalities that one is already familiar with, even if common sense tells one that one should modify them.

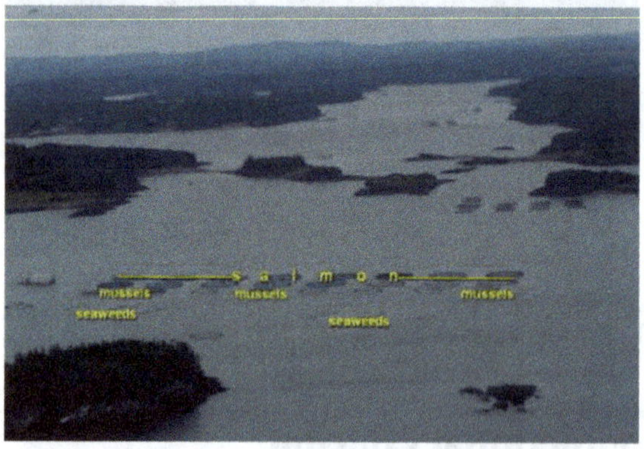

Aquaculture, Integrated Multi-trophic (IMTA). One of the Integrated Multi-Trophic Aquaculture (IMTA) sites in the Bay of Fundy, New Brunswick, Canada, operated by Cooke Aquaculture Inc.: two rows of salmon cages in the background, then a row of mussel rafts and two seaweed rafts in the foreground

Human society does not change quickly unless there are compelling reasons to do so. What to do when early large profit margins create short-term economic booms, followed within a few decades by dwindling meager profit margins? Often, the temptation is to throw more large volume cultivation operations and destructive methods into the mix, without proper regulations and business plans. Pollution, disease and economic busts generally ensure, major restructuring of the industry becomes necessary, and a few clairvoyant visionaries remain afloat and adapt to jump to the next curve to survive.

The fact that humans are currently at a crossroad should motivate them to improve current aquaculture practices, without further delay. Fishery management plans in most countries have been single-species approaches, completely neglecting the interactions between species, not understanding the synergies, or antagonisms, between them and how an ecosystem works based on the complementarities of the different functions of the different organisms inhabiting it. It seems that, despite the knowledge of the limitations of mono-agriculture and mono-fisheries, people are ready to develop similar plans for the management of mono-aquaculture. It should be recognized that there is still a chance for incorporating all the learning about the problems of terrestrial monocultures into the relatively new frontier of aquaculture.

To better manage marine, brackish, or freshwater environments to the benefits of mankind and the ecosystem, one needs to develop a new science, marine agronomy, learning from the mistakes made in land agriculture over the centuries to do a better job with aquaculture. It is interesting to note that traditional agricultural practices, such as crop alternation and fallow, are now being transposed to aquaculture practices. Why, then, is IMTA not more widely adopted, especially in the western world? Paul Greenberg, in his fascinating book "Four Fish", mentioned a very interesting point. In Leviticus, the third book of the Hebrew bible in which, according to the Jewish tradition, God dictated commandments to Moses, one can read (19:19): "You must not sow your field with two different kinds of seed" (also translated as "two kinds of seed" or as "mixed seed"). One can wonder if this represents, in fact, one of the most ancient treatises recommending mono-agricultural practices and if it is not the reason why integrated culture techniques have been ignored for centuries in the Judeo-Christian civilization, while they have flourished in other civilizations, especially in Asia. Moreover, if Asian cultures are accustomed to the concept of considering wastes from farming practices as resources for other crops rather than pollutants, this attitude still has a long way to progress in the western world where aquaculture is a more recent development.

The need for diversifying responsible aquaculture systems and for an ecosystem approach. The common old saying "Do not put all your eggs in one basket," which applies to agriculture and many other businesses, should also apply to aquaculture. Having excess production of a single species leaves a business vulnerable to sustainability issues because of fluctuating prices in what has become commodity markets and potential oversupply, and the possibility of catastrophic estruction of one's only crop (diseases, damaging weather conditions). Consequently, diversification of the aquaculture industry is advisable for reducing economic risk and maintaining sustainability and competitiveness.

From an ecological point of view, diversification also means cultivating more than one trophic level, i.e., not just raising several species of finfish (that would be "polyculture"), but adding into the mix organisms of different and lower trophic levels (e.g., seaweeds, shellfish, crustaceans, echinoderms, worms, bacteria, etc.) to mimic the functioning of natural ecosystems. Staying at the same ecological trophic level will not address some of the environmental issues because the system will remain unbalanced due to nondiversified input and output needs. Evolving aquaculture practices will require a conceptual shift toward understanding the working of food production systems rather than focusing on technological solutions.

One of the innovative solutions promoted for environmental sustainability (biomitigative services for improved ecosystem health), economic stability (improved output, lower costs, product diversification, risk reduction, and job creation in disadvantaged communities), and societal acceptability (better management practices, improved regulatory governance, and appreciation of differentiated and safe products) is IMTA. The aim is to increase long-term sustainability and profitability per cultivation unit (not per species in isolation as is done in monoculture), as some of the uneaten feed and wastes, nutrients, and by-products of one crop (fed animals) are not lost but recaptured and converted into fertilizer, feed, and energy for the other crops (extractive plants and animals). These, in turn, can be harvested and marketed as healthy seafood, while feed costs are reduced because of their reuse in multiple niches and biomitigation is taking place (partial removal of nutrients and CO_2, and supply of oxygen). In this way, all the cultivation components have a commercial value, as well as key roles in recycling processes and rendering biomitigative services. Some of the externalities of fed monoculture are internalized, hence increasing the overall sustainability, long-term profitability, and resilience of aquaculture farms. The harvesting of the different types of crops participates in the capture and export of nutrients outside of the coastal ecosystem.

The biomass and functions of the fed and extractive species naturally present in the ecosystem in which aquaculturefarms are operating must also be accounted for or this will lead to the development of erroneous carrying/assimilative capacity models. For example, the 158,811 t (fresh weight) of the intertidal seaweed, Ascophyllum nodosum (rockweed), in proximity to salmon aquaculture operations in southwest New Brunswick, Canada, are not neutral in the ecosystem and represent a significant coastal nutrient scrubber which should be taken into consideration to understand the functioning of that part of the Bay of Fundy.

IMTA as an Improvement Option for Aquaculture

IMTA has never been portrayed as the solution to and for everything! For example, IMTA does not address the issues of escapees from open-water fish farm s. It is, of course, in the interest of everybody, especially the industry (to not lose money) to reduce the number of escapees. This is, however, a question of engineering of the rearing systems (cages, netting material, etc.) and the suitability of the environment to survival should escapes occur. To solve the escapee issue, it has been suggested that fish farms should be pulled from the open water and placed on land or in closed containment. Moving on land is, however, not a guarantee for zero escapees.

There are well-known escapee cases from land-based operations, with serious consequences. For example, the bighead carp (Hypophthalmichthys nobilis) and the silver carp (Hypophthalmichthys molitrix) were brought from Asia to the southern USA in the 1970s to help control algal proliferation in channel catfish (Ictalurus punctatus) farms. There are reports of escapees into the lower Mississippi River system, especially associated with flood episodes in the early 1990s. Self-sustaining populations have been able to move northward to enter the Upper Mississippi River system and the Illinois River system. Presently, there are fears that these fish could enter the Great Lakes system through the Chicago Sanitary and Ship Canal and the Des Plaines River to finally reach Lake Michigan, after an escape of around 2,000 km in approximately 20-30 years. Electric fish barriers have been put in place, but their efficiency has been questioned. The use of rotenone, a biodegradable piscicide, was authorized but seemed to have killed more common carps (Cyprinus carpio; itself an introduced species from Europe in the 1830s) than bighead and silver carps. On April 26, 2010, the US Supreme Court decided not to get involved in a dispute over how to prevent these carps from making their way into the Great Lakes; it turned down a new request by the State of Michigan to consider ordering permanent closing of the Chicago-area shipping locks. What the impacts on the ecosystems could be, should these fish get into the Great Lakes systems, is unknown, but they are well-known for their ability to consume large amounts of algae and zooplankton, eating as much as 40% of their body weight per day, and they are fierce competitors when it comes to securing their food needs. The silver carp is also a danger to recreational fishers, water skiers, and boaters because of its habit to jump out of the water when startled by boat motors or other noises, creating life-threatening aerial hazards with high speed impacts.

The number of escapees from land-based facilities is not as well documented as with cage-based aquaculture. Perhaps because land-based fish escapes are more likely to occur as a continuous "trickle" instead of a single major event such as a net tear that would lead to "large-scale" escapes. However, reports do surface from time to time in the media, particularly if there is some novelty in the story. A recent example is the report of the cultured salmonid brown trout, Salmo trutta, escaping from a pond farm in the UK. A wildlife photographer caught them in action, making large leaps out of the water straight into a metal feed pipe a meter above and connected to a tributary of a river. Ideally, land-based recirculation systems would reduce the potential for escapes. However, most recirculation system s have at least partial water exchange and where there is water exchange and discharge, there is a potential for escapees. These systems are still not widely used and to the authors knowledge there has not been any initiative taken to document escapees, or lack thereof, within these systems. It may, therefore, be premature to classify such systems as "escape proof." It is unlikely that any land-based aquaculture operations could ever be 100% "escapee-proof" and consequently, they will also need to develop anti-escapee strategies (avoiding flood plains, electric fences, grids of the appropriate mesh, catchment basins, etc.).

Moving to land-based or closed-containment operations is one approach that may help address some sustainability issues but is not without its problems. Large amounts of energy, often diesel or electric power, are required to pump and aerate water. Nutrients are either pumped back into the water or settled somewhere and "trucked" offsite. All of these processes leave a "carbon footprint," and only partly solve the issue of excess nutrients. IMTA, or its variation called "aquaponics," will have to be added to closed-containment or land-based systems to treat the effluents. One "impact" may simply be traded for another. Ayer and Tyedmers, in their life cycle assessment of alternative aquaculture technologies, warned that one could be in a case of environmental problem shifting, not solving, where, while reducing local ecological impacts, the increase in material and energy demands may result in significant increased contributions to several environmental impacts of global concern, including global warming, nonrenewable resource depletion, and acidification.

Land-based or closed-containment operations have also been advocated as a way of controlling diseases and their transmission. However, the proponents very often equate diseases to the sole problem of sea lice, leaving the issues related to viral or bacterial pathogens unaddressed. Some concerns have been expressed that multiple species on the site might increase the risk for disease transmission. It must, however, be realized that sites in the ocean and on land will always have additional unintended species associated with the operation, ranging from microorganisms to marine mammals, depending on the situation. The question is not whether to have only one species on the site, but at what density do negative interactions occur with the unintended ones and whether there are any positive interactions associated with more diversified systems. In fact, two studies have demonstrated in laboratory experiments that the blue mussel, Mytilus edulis, is capable of inactivating the infectious salmon anemia virus (ISAV), as well as the infectious pancreatic necrosis virus (IPNV). Mussels are, consequently, not a likely reservoir host or vector for ISAV and IPNV. Put in an IMTA perspective, this could mean that mussel rafts could be strategically placed to serve as a kind of sanitary/biosecurity cordon around salmon cages to combat certain diseases. Pang et al. also reported reduced total bacteria and Vibrio counts in a seaweed-abalone IMTA system.

In regard to parasites, two studies [14; Robinson, pers. comm.] indicate that blue mussels can consume copepodids, the planktonic and infectious stage of sea lice, and several studies, in both Europe and New Zealand, have highlighted the fact that mussels can consume small zooplankton. Having a biofilter such as mussel s at IMTA sites may decrease the frequency of exposure to pathogens and planktonic parasites. The hope is that having multiple species on a farm will result in some positive interactions between species allowing some biological control of the outbreaks of pathogens and parasites, hence reducing the number of costly chemical treatments required. If this is validated, filter feeders may have additional contributions to sustainability beyond reduction of the particle load. One of the 14 projects of the recently created Canadian Integrated Multi-Trophic Aquaculture Network (CIMTAN) is investigating the role of bivalves in potentially reducing sea lice populations. Most of the work has been conducted in the lab so far, but results are very positive and it has been demonstrated that mussels eat the larval forms. Ongoing work is developing a trap system that exploits various behaviors of sea lice to attract and filter them out of the system. Another CIMTAN project is looking into the possibility that mussels could reduce the horizontal transmission of Loma salmonae, responsible for microsporidial gill disease of salmon (MGDS), a serious endemic gill disorder in marine netpen reared, and wild, Chinook (and other Pacific) salmon. Trials will examine the proof of principle that blue mussels remove microsporidial spores from water and to what extent these spores retain short-term infectious potential as determined by branchial xenoma expression in test fish.

IMTA is not entering directly the debate regarding the inclusion of fishmeal and fish oil in commercial feeds (nor are land-based or closed-containment operations). IMTA could, however, provide a partial solution. Modern commercial salmon diets in Canada contain much less fishmeal (about 15-25%) and fish oil (about 15-20%) than they did less than 10 years ago (40-60%). By-products (trimmings, offal) of wild catch fisheries are now used to supply a major portion of the fishmeal ingredients. Finding replacements for marine ingredients is a priority and there are several large research projects worldwide addressing this issue. The feed company Skretting has now produced a salmon feed which includes no marine ingredients. Turning toward land plant proteins is not without its impacts. Extra farmland area (more deforestation) would be needed, which, moreover, would need to be irrigated and fertilized on a planet already suffering from water availability problems and with fertilizer prices soaring. The price of some staple food crops used in traditional agriculture (corn, soya bean, sugar cane, etc.) would rise considerably due to announced competition for their uses, as recently seen when they were potentially sought out as energy crops for the production of first-generation biofuel s. Reallocation of acreage for subsidized potential biofuel crops such as corn, sugar cane, oil palm, canola, switch grass, etc., has already had significant ecological and societal costs due to its impacts on ecosystem health, biodiversity, and food security. Partial substitution with organisms already living in water and not needing extra fertilization in an IMTA setting, such as seaweeds, could, in fact, be a very interesting option, fitting well within the sustainability and management concept of IMTA, and representing a logical loop for companies developing an IMTA and diversification strategy. If cultivated in the water column in IMTA systems, there would, moreover, be no issue of raking natural beds of seaweeds attached to the bottom of the ocean (destruction of seafloor and impact on ecosystem functions such as nursery ground for animals).

Some environmental nongovernmental organizations arguing for fishmeal/fish oil replacement have also voiced concerns that, after all, marine fish should eat marine ingredients obviously, one cannot have it both ways. There is also the paradoxical situation of farmed freshwater fish, which are being grown less and less on humans and animal wastes and naturally occurring algal blooms, but more and more on already competing staple foods such as corn and soy: they have lost their off-flavored or muddy taste to become tasteless or “unfishy”. So, what does one want to receive in one’s kitchen? A flavor-neutral, versatile product easily adapted with numerous sauces, while one is lamenting that farmed salmon or bass are not what wild salmon or bass used to be? Quite an irony, even more so when people learn that these herbivorous whitefish are more and more being fed pellets containing fishmeal and fish oil because they grow faster. What is really important is a balanced diet using balanced sourcing of raw material.

Some will argue that “fish require nutrients, not ingredients.” At the same time, there is also the well-known saying “You are what you eat,” and in this case, people have to realize and accept that humans are mostly corn, soya, and fishmeal, if they look at what the four mammals (cow, pig, sheep, and goat) and four poultry (chicken, turkey, duck, and goose) that they have selected as their meat choices are eating. Historically, most of the reduction fishery (small fish such as anchovies, herring, sardines, and menhaden) went into the production of pet feeds and farm animal feeds. Subsequently, this fishery supplied a significant part of the marine ingredients for fish feeds. The landing of the reduction fishery has been fairly stable (fluctuating between 15 and 30 million metric tons since the 1970s) and, in the absence of aquaculture, the fishery would likely return to supplying pet and farm animal feeds, and a current resurgence of interest directly by humans. This is not to justify relaxed vigilance for finding replacements for marine ingredients in fish feed, but simply to suggest that an absence of fish farming will not stop the use of this resource. How can one get out of this vicious circle? Cultivating several organisms, at different trophic levels, in proximity so that the food and wastes are utilized efficiently more than once through a cascade of recapturing and remetabolizing is one approach: that is IMTA. The other is to consider that if terrestrial food production systems are close to their limits, one does not have other options but to turn again to the sea, this time not for fish but to have seaweeds and invertebrates entering one's food habits, either directly or delivered through feed given to intermediates to what reach one's plate. The discrepancy between the marine and agricultural production systems has to be reduced: presently, especially in the western world, humans feed approximately two steps higher in the marine food web than in the agricultural food web.

People should continue to eat seafood (fish but also invertebrates and seaweeds), not according to seafood pocket guides which simplistically paint species with one stroke of green (best choice), orange (good alternative), or red (avoid), but according to the fishing and aquaculture practices used to grow, harvest, and process them: an admittedly more complex mosaic, but also much more realistic and attractive to look at than a traffic light! Interestingly, what is referred to as the fifth tasting sense by Japanese (after sweet, sour, salty, and bitter) and called umami (= savoriness or good flavor) comes from seaweeds. The product responsible for umami was first identified in 1908 by Professor Kikunae Ikeda, of the Tokyo Imperial University, searching for the chemical reason of the strong flavor in seaweed broth (mostly of the kelp Saccharina japonica, formerly Laminaria japonica). It is due to the detection in our mouth of the carboxylate anion of the amino acid called glutamic acid and its salts, glutamates, in particular monosodium glutamate (MSG). Inosine monophosphate (IMP) and guanosine monophosphate (GMP), degradation products of the energy-storing molecule adenosine triphosphate (ATP) greatly enhance the perceived intensity of umami. This explains, chemistry displacing romantics, why a dead tuna (once full of energy) served with seaweeds is such a savory delicacy, the very essence of the success of the sushi bar fad gaining the western world.

We have never pretended that IMTA is the solution, the silver bullet, to and for everything. It is now up to us to develop the better aquaculture practices of tomorrow. IMTA is based on several common sense principles:

- The solution to nutrification is not dilution, but extraction and conversion through diversification.
- This is, in fact, a rewording of the first law of thermodynamics "Rien ne se perd, rien ne se crée, tout se transforme" ("Nothing is lost, nothing is created, everything is transformed") as summarized by Antoine Laurent de Lavoisier, the well-known French chemist and physicist (but also tax collector, which explains his premature death at age 51 in 1794 under the Terror period of the French Revolution).

- What is waste for some is gold for others.
- Do not put all your salmon eggs in the same basket.

A lot of common sense, but, unfortunately, common sense is not that common! IMTA is one of the promising options, but, certainly, it needs to be tailored to the location in the world where it is implemented. It should also be developed in association with other practices. Like for energy, not one solution will satisfy all the needs and it is a variety of solutions that will help one secure one's seafood procurement in a responsible manner. The solutions will be at the interfaces of these techniques and will be interdisciplinary. They will embrace both scientific and technological advancements and traditional knowledge. IMTA is exactly at this interface, modernizing traditional practices: combining ecosystem complementary crops, bay management area, and fallowing are nothing new, but revisited and updated, based on what humans have learned from past experience (which includes a lot of mistakes over the centuries, but not assimilated by the characteristically short-term memory of humans).

A few economic analyses have indicated that the outlook for increased profitability through IMTA is promising. However, these analyses were based solely on the commercial values from the sale of biomass - being of fish, shellfish, or seaweeds - and used conservative price estimates for the cocultivated organisms based on known applications. One aspect not factored into these analyses is the fact that the extractive component of an IMTA system not only produces a valuable multipurpose biomass, but also simultaneously renders waste reduction services to society. It is particularly important to recognize that once nutrients have entered coastal ecosystems, there are not many removal options available: the use of extractive species is one of the few realistic and cost-effective options. The economic values of the environmental/societal services of extractive species should, therefore, be recognized and accounted for in the evaluation of the true value of the IMTA components. Further development of economic models is needed to help shed light on the economic (society) and commercial (industry) attractiveness of IMTA.

Ecosystem services have been ignored until recently. To improve the sustainability of anthropogenic nutrient loading practices such as aquaculture, incentives such as Nutrient Trading Credits (NTC) should be established as a means to promote nutrient load reduction or nutrient recovery. During the last few years, there has been much talk and excitement about carbon credits. However, within coastal settings, the concerns have largely been with nitrogen, due to the fact that its typical role as the limiting nutrient is not any longer the case in some regions. Potential effects of carbon loading in the marine environment should also be considered: localized benthic anoxia and, consequently, hydrogen sulfide release may occur when solid waste deposition rate exceeds aerobic decomposition rate. Ocean acidification due to increased dissolved CO_2 levels has also prompted serious new concerns and a Carbon Trading Credit (CTC) system should also be contemplated. With an appropriate composition of cocultured species, IMTA has the potential to reduce the amounts of dissolved (inorganic) and solid (organic) forms of nitrogen, carbon, phosphorus (more an issue in freshwater environments), etc., making extractive aquaculture a good candidate for a NTC and CTC, or other suitable approaches, to deal with the pressing issues of coastal nutrient loading.

Currently, there are few countries with laws or regulations that require aquaculture operations to responsibly internalize their environmental costs, such as nutrient discharges. There are some precedents, such as where land-based trout farmers in Denmark are allowed to increase their feed quota with documented evidence of reduced effluent discharge, but such incentives are not widely spread. In most jurisdictions, adjacent ecosystems are left to accommodate the nutrient load, and performance-based standards are used to determine if farms have exceeded their assimilative capacity.

The implementation of regulations resulting in internalization of environment costs by fish farms, without a direct economic compensatory response such as the Danish feed quota increase, could result in a significant reduction in profitability. In land-based systems, it is relatively easy to quantify nutrient load and concentration via comparison between farm inflows and outflows, thereby creating a benchmark for "economic compensation." Such values arepractically impossible to empirically measure in an open-water system, "leaky" by definition, and, consequently, so is the practical implementation of such incentives. However, Troell et al. and Chopin et al. demonstrated that by integrating the seaweed, Gracilaria, in the dual role of nutrient scrubber and commercial crop (for agar production), with salmon farms in Chile, the environmental costs of waste discharges would be significantly reduced and profitability significantly increased.

Interestingly, the removal of nitrogen could be much more lucrative, by approximately a factor of 100, than that of carbon. The cost of removing nitrogen is not clearly defined, but there are several interesting studies that may help define a range of possible prices for economic evaluation of the NTC concept. The cost of removing 1 kg of nitrogen varies between US$3 and US$38 at sewage treatment facilities, depending on the technology used and the labor costs in different countries. The municipality of Lysekil, in Sweden, is paying approximately US$10/kg removed by the filter-feeding mussel, Mytilus edulis, to the farm Nordic Shell Produktion AB. Ferreira et al., with the development of the Farm Aquaculture Resource Management (FARM) model, determined a net value of €18-26 billion/year of nutrient eutrophication reduction services provided by shellfish aquaculture in the coastal waters of the European Union. Gren etal. calculated that the cleaning costs of nutrients by mussel farming can be considerably lower than other abatement measures and estimated that mussel farming should be credited between €0.1 and €1.1 billion/year in the Baltic Sea. Using this information, and without presuming what the final design of IMTA sites will be in the future, preliminary calculations for the relatively small-scale IMTA project on the East coast of Canada indicate that the annual harvesting of kelps would equate to the removal of 35.75 t of nitrogen from the ecosystem, representing an NTC of between US$357,504 and US$1,072, 512. The same could be applied to another key nutrient, phosphorus. With an annual removal of 4.09 t and a value of US$4/kg removed, this would represent another contribution to the NTC of US$16,343, a much smaller amount but it could also be an important way of extracting phosphorus, at a time when some are predicting it to be the next element human society will be short of (in its natural or mined forms).

Carbon Trading Credits (CTC) could also be calculated. There may be some arguments about what is meant by trapping and sequestering carbon. Some may argue that it should be reserved to long/geological term storage (sink) and not to transient storage. This is, in fact, a question of how long one allows the recycling clock to run. There is no permanent storage of carbon; it happened that a particular fossil biofuel, petroleum, has been sequestered over geological time to suddenly be reused at an accelerated rate over the last few centuries. But the first law of thermodynamics, as enunciated by Antoine Laurent de Lavoisier more than two centuries ago, still applies: "Rien ne se perd, rien ne se crée, tout se transforme," i.e., "Nothing is lost, nothing is created, everything is transformed." If even temporary removal of carbon from the ocean by biomass harvesting until further transformation (and rerelease of carbon) can be credited for potentially increasing seawater pH and absorbing CO_2 from the atmosphere and/or the cultivated animals, then CTC should be calculated. Marine vegetation is getting more and more recognition as a sink for anthropogenic carbon emissions (the so-called blue carbon).

Aquaculture, Integrated Multi-trophic (IMTA). Harvesting of the kelp, Saccharina latissima, at an Integrated Multi-Trophic Aquaculture (IMTA) site in the Bay of Fundy, New Brunswick, Canada. Kelps remove dissolved nutrients from the ecosystem while providing commercial products

Marine primary producers contribute at least 50% of the world's carbon fixation and may account for as much as 71% of all carbon storage in oceanic sediments. Then, micro-alga e, macro-algae, and marine plants, such as mangroves and seagrasses, have a role to play in CO_2 sequestration and removal, and carbon storage. Marine photosynthesis accounts for 50% of the total primary productivity of the planet (54-59 PgC/year from a total of 111-117 PgC/year). Of this, marine macrophytes (seaweeds and seagrasses) account for approximately 1 PgC/year concentrated in coastal regions where they can play a significant role in the sequestration of anthropogenic carbon emissions and the global carbon cycle. Brown marine macro-algae (such as Macrocystis, Saccharina, Laminaria, Ecklonia, Sargassum, Ascophyllum, and Fucus), red algae (such as Porphyra, Palmaria, Eucheuma and Gracilaria) and green algae (such as Ulva), are capable of very high rates of photosynthesis and productivity. These rates of productivity compare very favorably to those of terrestrial crops that have been recommended as possible sources of first-generation biofuels (corn, Zea mays) or second-generation biofuels (switch grass, Panicum virgatus; E-grass, Miscanthus giganteus) and position marine macro-algae very well for being part of the third-generation biofuels.

Coming back to the IMTA project on the East coast of Canada, using a value for carbon removal of around US$30/t, the annual harvesting of kelps would represent an annual removal of 306.43 t and a CTC of US$9,193: a larger amount of carbon, but for a much smaller value of trading credits, underlining the difficulty in removing dissolved nutrients from aquatic systems and the acute issue of their presence in coastal systems. Similar calculations could be applied to the organic extractive component of IMTA. In the case of shellfish, accumulation of nitrogen, phosphorus, and carbon should be considered both in meat and shells, which are especially rich in calcium carbonates.

At a much larger scale, the occurrence of large and recurrent "green tides" should also be brought into focus. Large proliferations of opportunistic green algae, especially of the genus Ulva, in response to large anthropogenic nutrient loading, have been in the news over the last few years in places around the world such as Northern Brittany in France, the southern regions of the UK, and

Venice in Italy. The green tide in Qingdao, China, just before the sailing competitions of the 2008 Olympic Games, got a lot of attention. The following question needs to be asked: Are these green tides a negative media photo opportunity, or are they reminders of the significant role seaweeds play in coastal processes and the services they render? Within 3 weeks, 1 million tons of Ulva prolifera were removed from the vicinity of Qingdao to allow the sailors and windsurfers to compete (but it is estimated that approximately 2 million tons of U. prolifera sank to the bottom of the Bay; another environmental problem shifting, but not a solution). The harvesting of 1 million tons equated to between 3,000 and 5,000 t of nitrogen removal for a NTC value between US$30 and US$150 millions. Additional NTC of US$1.6 million for the removal of 400 t of phosphorus, and CTC of US$900,000 for the removal of 30,000 t of carbon, should also be factored in.

Aquaculture, Integrated Multi-trophic (IMTA). A green tide of Ulva prolifera in Qingdao.

A smaller green tide occurred in 2007. Large ones were also reported in 2009 and 2010 but they stayed offshore in the Yellow Sea. Out of sight should, however, not mean out of mind. If urgent measures are not taken, this will be a recurrent event for years to come. Is there a solution? Green tides are not the cause, but the unintentional consequence of coastal eutrophication. With the presence of sufficient nutrients and solar energy, these opportunistic species, with a well-adapted anatomy, morphology, and physiology, will proliferate. Obviously, it would be beneficial to reduce nutrient loading at the source, but this may not be possible in the present context of economic development along China's coastal zone. The problem is that U. prolifera is presently an unwanted and uncontrolled growing nuisance species of limited commercial value. To control its proliferation, the solution may be to create a competition for nutrients by intentionally cultivating algal species, which not only carry on the biomitigation, but also have a commercial value, where U. prolifera starts to enter the coastal environment (discharges from juvenile river crab land-based aquaculture ponds along Jiangsu province, south of Shandong province where Qingdao is located). This time, the IMTA concept has to be interpreted as an integrated land pond/coastal aquaculture system in a supra Integrated Coastal Zone Management (ICZM) effort, beyond provincial borders, to address issues at the Yellow Sea scale. It is understood that this "out of the box" approach to ICZM will, initially, raise eyebrows as the idea of growing more seaweeds (but of commercial value) to contain the proliferation of other seaweeds, presently considered nuisances, is not the most intuitive approach for a lot of people or decision makers.

At the present time, there seems to be a stage of recognition, awareness, and communication of the concepts of ecosystem services and biomitigative services rendered by extractive aquaculture. Next will come the time to transform the concepts into biomitigative solutions and then their inclusion in regulatory and management frameworks. Establishing and implementing a structure for the payment schemes (credits or incentives) of these services will be a delicate matter. Will it be one agency, but with funds coming from where? Should it be a regional, national, or international agency(ies), trading at which scale(s)? Will an extractive aquaculture operation in existence for many years receive credits, or will only the new ones? Would a fed aquaculture operation also practicing extractive aquaculture be eligible for credits, or will it be the case for the extractive only aquaculture operations? What about the situation in which people run both types of farms. Moreover, due to complex hydrographic and current patterns, it is obvious that extractive species at a site are not limited to absorbing/sequestering the nutrients generated exclusively at that site. Consequently, is it possible to establish a clear spatial nutrient removal budget which would be associated with the corresponding credits/incentives? Will the sequestration have to be "permanent," or will a temporary removal/storage be acceptable and more realistic? A lot of regulatory details will have to be worked out before this complex scheme becomes a reality.

What Will it Take to Increase the Acceptance and Adoption of IMTA as a Responsible Aquaculture Practice of the Future? Presently, the most advanced IMTA systems in open marine waters and land-based operations have three components (fish, suspension feeders or grazers such as shellfish, and seaweeds, in cages, rafts, or floating lines), but they are admittedly simplified systems. More advanced systems will have several other components (e.g., crustaceans in mid-water reefs; deposit feeders such as sea cucumbers, sea urchins and polychaetes in bottom cages or suspended trays; and bottom-dwelling fish in bottom cages) to perform either different or similar functions, but for various size ranges of particles, or selected for their presence at different times of the year (e.g., different species of seaweeds). The most advanced IMTA systems, near or at commercial scale, can be found in Canada, Chile, South Africa, Israel, and China. Ongoing research projects related to the development of IMTA are taking place in the UK (mostly Scotland), Ireland, Spain, Portugal, France, Turkey, Norway, Japan, Korea, Thailand, the USA, and Mexico. It will also be interesting to observe how new seaweed cultivation initiatives in different parts of the world for biofuel production could be an additional driver to adopt IMTA practices.

For IMTA to develop to a commercial scale, appropriate regulatory and policy frameworks need to be put in place. Present aquaculture regulations and policies are often inherited from previous fishery frameworks and reasoning, which have shown their limitations. It is, therefore possible that some of the existing regulations and policies could impose unintentional constraints on the future growth of IMTA. To develop the aquaculture of tomorrow, current governance structures pertaining to aquaculture need to be revisited and reviewed with the aim of identifying changes in the regulatory/policy environment that are needed to facilitate the operation of IMTA farms. Adaptive regulations need to be developed by regulators with flexible and innovative minds, who are not afraid to put in place mechanisms that allow the testing of innovative practices at the R&D level, and, if deemed promising, mechanisms that will take these practices all the way to C (commercialization). As the IMTA concept continues to evolve, it is important that all sectors of the industry are aware of the implications of the changes involved, so that they can adapt in a timely and organized manner.

To move research from the "pilot" scale to the "scale up" stage, some current regulations and policies may need to be changed or they will be seen as impediments by industrial partners who will see no incentive in developing IMTA. For example, an earlier version of the Canadian Shellfish Sanitation Program (CSSP) prevented the development of IMTA because of a clause that specified that shellfish could not be grown closer than 125 m of finfish netpens. This paragraph was clearly not written with IMTA in mind, but it seriously impinged its development. After 4 years (2004-2008), it was amended so that IMTA practices could develop to commercial scale legally, based on recent, reliable, and relevant data and information provided by three government departments and the IMTA project on the east coast of Canada. While 4 years may seem long, it is a relatively short delay considering that regulations and legislations require thorough review with due governmental process involving several federal and provincial departments. This suggests that new aquaculture practices should be accompanied by timely regulatory review to avoid market delays for new products. As governments move to revise current regulatory regimes, it will be necessary to press the importance of accommodating and indeed encouraging new sustainable solutions such as IMTA. IMTA also requires approaching aquaculture development and management with a holistic approach and not one species, or group of species, at a time. It is known that this approach has led to many failures in the management of the fisheries; vigilance is required so that the same flaw is not repeated in the management of aquaculture.

Most current aquaculture business models do not consider or recognize the economic value of the biomitigative services provided by biofilter s, as there is often no cost associated with aquaculture discharges/effluents in land-based or open-water systems. In order to ensure further development of IMTA systems worldwide, from the experimental concept to the full commercial scale, defining and implementing the appropriate regulatory and policy frameworks, and financial incentive tools such as NTC and CTC, may therefore be required to clearly recognize the benefits of the extractive components of IMTA systems. Better estimates of the overall costs and benefits to nature and society of aquaculture waste and its mitigation would create powerful financial and regulatory incentives to governments and the industry to jointly invest in the IMTA approach, as the economic demonstration of its validity would be even more obvious. Moreover, by implementing better management practices, the aquaculture industry should increase its societal acceptability, a variable to which it is very difficult to give a monetary value, but an imperative condition for the development of its full potential. Reducing environmental and economic risk in the long term should also make financing easier to obtain from banking institutions.

The determination to develop IMTA systems will, however, only come about if there are some visionary changes in political, social, and economic reasoning. This will be accomplished by seeking sustainability, long-term profitability, and responsible management of coastal waters. There is still a large amount of education required to bring society into the mindset of incorporating IMTA into their suite of social values. Some of the attitudinal surveys conducted in Canada and the USA indicate that the general public is in favor of practices based on the "recycling concept." Consumers' perceptions and attitudes may also have to change. Why is recycling and the concept of "what is waste for some is gold for others" well accepted in agricultural practices, but is not yet acquired when transposed to aquaculture practices? Will consumers come to accept eating products cultured in the marine environment in the same way they accept eating products from recycling and organic agricultural practices, for which they are willing to pay a higher price for the perceived higher quality or ethical premiums? After all, regulations require mushrooms to be specifically grown on farmyard manure and animal excrements to receive organic certification. Will a greater appreciation of the sustainable ecological value of the IMTA concept, a willingness to support it tangibly with shopping money, and an increased pressure on elected representatives emerge? This will be the ultimate test. The degree to which researchers and extension people become creatively involved with this educational component will be vital to the success of IMTA practices. The differentiation of IMTA products through traceability and eco-labeling will also be key for their recognition and command of premium market prices.

Some have argued that the adoption of IMTA in the western world is slow. For example, on the east coast of Canada, there were obviously no IMTA sites in the Bay of Fundy in 2001 when IMTA research started. Nine years later, 8 of the 96 finfish sites in southwest New Brunswick have the combination salmon (or cod)/mussels/kelps and 8 other sites have been amended to develop IMTA. This is a respectable conversion of almost 16% in 9 years. Moreover, it would not be reasonable to anticipate an instant conversion, as the industry needs to develop markets to absorb the cocultured biomass: this also takes time and can only be progressive.

Future Directions: The Path Forward

Several IMTA projects, worldwide, have now accumulated enough data to support the proof of concept at the biological level. The next step is the scaling up of more experimental systems to commercial scale to further document the economic and social advantages of the concept, which will be key to offering IMTA to practitioners of monospecific aquaculture as a viable option to their current practices. Emerging responsible aquaculture approaches must generate net economic benefits for society if they are to be advocated. Working on appropriate food safety regulatory and policy frameworks in the respective countries will be essential for enabling the development of commercial scale IMTA operations in a more universal fashion.

It has taken decades to reach current finfish aquaculture production levels and learn new species husbandry. A major rethinking is, however, needed regarding the definition of an "aquaculture farm " by reinterpreting the notion of site-lease areas and regarding how it works within an ecosystem, in the context of a broader framework. Within Integrated Coastal Zone Management (ICZM), integration can range from the small scale (a leased site with its spatial limits) to a Bay Management Area (BMA) and to the larger scale of a region connected by the functionalities of the ecosystem. Amending regulations to allow a new type of aquaculture systems will not occur overnight. This should, however, not discourage the finfish aquaculture industry from practicing IMTA, as even small amounts of cocultured species production are useful at the initial stage of development.

Selecting the right combination of species will be critical. They will have to be appropriate for the habitat, the available culture technologies and labor forces, and the environmental, climatic, and oceanographic conditions. They will have to be complementary in their ecosystem functions, growing to a significant biomass for efficient biomitigation, commanding an interesting price as raw material or presenting an interesting added value for their derived products. Their ecological interactions and synergies within an IMTA system will have to be identified and understood to take full advantage of them. Their commercialization should not generate insurmountable regulatory hurdles.

Optimal design will not only facilitate nutrient recovery, but should also promote augmented growth beyond what would be expected were these species cultured in isolation. In addition to the obvious economic return from increased growth rates from additional species, some less tangible benefits should also be factored in, such as the biomitigative services rendered by the extractive species. Economic analyses will have to recognize and account for the values of the environmental/societal services of extractive crops to estimate the true value of these IMTA components. Economic analyses will need to be part of the overall modelling of IMTA systems, as they get closer to commercial scale and their economic benefits and costs, as well as their impacts on coastal communities, are better understood. It will then be possible to add profitability, resilience, social/economic desirability, and economic impacts to the comparison between IMTA and monoculture settings. They will have to include the pricing and marketing potential and impact of organic and other eco-labelling s, the value of biomitigative services for enhanced ecosystem resilience, the savings due to multi-trophic conversion of feed and energy which would otherwise be lost, and the reduction of risks through crop diversification and increased societal acceptability of aquaculture (including food safety, food security, and consumer attitudes toward buying sustainable seafood products).

This would create economic incentives to encourage aquaculturists to further develop and implement sustainable marine agronomy practices such as IMTA, and would increase the societal acceptability of aquaculture by the general public. Seaweeds and invertebrates produced in IMTA systems should be considered as candidates for a variety of regulatory measures that internalize these benefits. For example, nutrient and carbon trading credits (NTC and CTC) could be used to promote nutrient removal, CO_2 sequestration, oxygen provision, and coastal eutrophication reduction within the broader context of ecosystem goods and services. Long-term planning/zoning promoting biomitigative solutions, such as IMTA, should become an integral part of coastal regulatory and management frameworks.

Nutrient extractive aquaculture appears to be a viable ecological engineering option for managing/internalizing some of the externalities generated by aquaculture operations. Effective government legislation/regulations and incentives to facilitate the development of IMTA practices and the commercialization of IMTA products will be necessary. The development and adoption of technology often depends in part on the level of legislative pressure from a nation's government, itself reacting to pressures from consumers, environmental nongovernmental organizations, and the public at large. If environmental legislation remains a low priority with government, then little progress toward the use of biofilters (as a means of effluent mitigation) will occur. The only motivator will be profits obtained from additional product growth and regulatory incentives. Therefore, if governments put legislative pressure on the proper management of wastewater effluent, openly support the use of biomitigation for effluent management, and put in place the appropriate corresponding financial tools (funding for IMTA Research & Development, outreach and technology transfer, and NTC and CTC incentives), then the development of IMTA will be encouraged.

Permissions

All chapters in this book are published with permission under the Creative Commons Attribution Share Alike License or equivalent. Every chapter published in this book has been scrutinized by our experts. Their significance has been extensively debated. The topics covered herein carry significant information for a comprehensive understanding. They may even be implemented as practical applications or may be referred to as a beginning point for further studies.

We would like to thank the editorial team for lending their expertise to make the book truly unique. They have played a crucial role in the development of this book. Without their invaluable contributions this book wouldn't have been possible. They have made vital efforts to compile up to date information on the varied aspects of this subject to make this book a valuable addition to the collection of many professionals and students.

This book was conceptualized with the vision of imparting up-to-date and integrated information in this field. To ensure the same, a matchless editorial board was set up. Every individual on the board went through rigorous rounds of assessment to prove their worth. After which they invested a large part of their time researching and compiling the most relevant data for our readers.

The editorial board has been involved in producing this book since its inception. They have spent rigorous hours researching and exploring the diverse topics which have resulted in the successful publishing of this book. They have passed on their knowledge of decades through this book. To expedite this challenging task, the publisher supported the team at every step. A small team of assistant editors was also appointed to further simplify the editing procedure and attain best results for the readers.

Apart from the editorial board, the designing team has also invested a significant amount of their time in understanding the subject and creating the most relevant covers. They scrutinized every image to scout for the most suitable representation of the subject and create an appropriate cover for the book.

The publishing team has been an ardent support to the editorial, designing and production team. Their endless efforts to recruit the best for this project, has resulted in the accomplishment of this book. They are a veteran in the field of academics and their pool of knowledge is as vast as their experience in printing. Their expertise and guidance has proved useful at every step. Their uncompromising quality standards have made this book an exceptional effort. Their encouragement from time to time has been an inspiration for everyone.

The publisher and the editorial board hope that this book will prove to be a valuable piece of knowledge for students, practitioners and scholars across the globe.

Index

www.ingramcontent.com/pod-product-compliance
Lightning Source LLC
LaVergne TN
LVHW080651210926
848485LV00003B/359